Fission Damage
in Crystals

Fission Damage
in Crystals

LEWIS T. CHADDERTON

Science Center/Aerospace and Systems Group,
North American Rockwell Corporation,
Thousand Oaks, California, USA

and

IAN McC. TORRENS

Centre d'Études Nucléaires de Saclay,
Gif-sur-Yvette, France

METHUEN & CO LTD
11 NEW FETTER LANE · LONDON EC4

First published 1969
© 1969 *Lewis T. Chadderton and Ian McC. Torrens*
Printed in Great Britain by
Spottiswoode, Ballantyne and Co. Ltd.,
London and Colchester
SBN 416 12420 8

Distribution in the U.S.A.
by Barnes & Noble, Inc.

Preface

In an earlier monograph (*Radiation Damage in Crystals*, Lewis T. Chadderton, Methuen & Co Ltd, 1965) an account has been given of the fundamentals of the physics of radiation damage. In the present book we deal more closely with the specific problem of understanding the damaging processes which occur in a single crystal during and immediately following the onset of fission. Furthermore, we neglect consideration of the damage produced by the fission neutrons, studies of which are reported in other research monographs, and focus attention on the heavy, energetic atomic fragments which emerge from a fission event.

The distinguishing features of the typical fission fragment are the high mass, energy, and charge which it possesses at birth. For this reason it is a highly efficient particle for the production of dense concentrations of point defects and dislocations. Yet, because of these unusual properties, the mechanisms by which damage is generated are highly complex and are not at all well understood. This book therefore describes an essentially experimental approach which employs direct atomistic procedures – electron and field-ion microscopy, computer simulation, etc – and deals mainly with the work of the authors and their collaborators. No discussion is made of the damage produced by fission particles in polycrystalline films or of post-irradiation etching experiments; though excellent reviews are available elsewhere (Allen N. Goland, *Studies in Radiation Effects*, edited by G. J. Dienes, p. 159, Gordon and Breach, 1966).

The early part of the research programme was directed by Professor F. P. Bowden, CBE, FRS, in the Surface Physics Laboratory, Cambridge, England, and much of the later work was performed by our colleagues working in the same laboratory. Our indebtedness to others is accordingly very great and it is a privilege to be able to report their work in this more complete form. We are particularly indebted to Professor Bowden and Dr A. D. Yoffe, and to Drs K. M. Bowkett, P. E. Caspar, A. G. Fitzgerald, J. Greenwood, M. Mannami, H. M. Montagu-Pollock, D. V. Morgan, H. Nordén, D. Norris, B. Ralph, M. J. Sole, and D. Van Vliet, in Cambridge. In California we have received unswerving assistance from our photographer, P. K. Ocorr, from our artist, F. S. Fegette, and from our most efficient secretary, Joan Harman. The book is dedicated to those who have made it possible.

LEWIS T. CHADDERTON
IAN MCC. TORRENS

Thousand Oaks
California
January, 1968

vii

Contents

CHAPTER ONE

Nuclear Fission

The Atomic Nucleus

The extended sequence of investigations which led to the discovery of nuclear fission was initiated in 1896 by the French scientist Henri Becquerel. While studying the phosphorescence of uranium salts he observed that nearby photographic emulsions were darkened just as much when carefully shielded for long periods of time from excitation by incident light as they were when completely exposed. He concluded that there was some *other* form of emanation from the material itself.

The nature of the emanation was not understood, however, for several years, although the physics of the *radioactivity*, as it was subsequently called, was extensively studied. One significant observation was that a radioactive substance could, in certain cases, induce radioactivity in a different substance placed nearby. This, and many other results which emerged from the study of radioactive materials, led to the inevitable conclusion by Rutherford and Soddy (1903) that the atoms of an element were not the fundamental building blocks of nature which had served so excellently the previous century's scientists, but that some heavy atomic species were subject to radioactive disintegration and consequent transmutation into other species, with the simultaneous ejection of light particles.

The next major step, which changed completely our conception of the structure of matter, was the hypothesis of the nuclear atom by Lord Rutherford in 1911. Rutherford deduced, from the results of experiments concerned with the scattering of α-particles by thin metal foils, that the large angular deflections observed were the result of an electrostatic interaction between the positive charge of the α-particle and that of a concentrated positively charged nucleus of diameter less than 10^{-12} cm. Since the α-particles used in the scattering experiments were themselves emitted by radioactive nuclei (Gamow, 1928), we might with impunity observe that the nucleus was instrumental in its own discovery. Moreover, the fact that the diameter of a nucleus was some four orders of magnitude less than that of the atom implied that matter was very largely empty space – in fact greater than 99·9% of the mass of an atom was concentrated in the nucleus, yielding a density there of some 100 million ton/cm^3.

1

Further scattering experiments with foils of different metals suggested an empirical rule for the radius r of a nucleus, which related it to the mass number A:

$$r = r_0 A^{1/3} \qquad (1.1)$$

The constant, r_0, the radius of the nucleus of the hydrogen atom (the proton), was of the order of $1 \cdot 2 \times 10^{-13}$ cm. This relationship led to the belief that a nucleus of mass number A and atomic number Z consisted of A tightly bound protons, and $(A - Z)$ electrons to account for the positive charge Z; a hypothesis which persisted until the discovery of the neutron by Chadwick in 1932. One attraction of this proton-electron theory was the possibility of explaining β-radioactivity, since β-particles were identified as electrons. However, with the advent of quantum mechanics, several objections to the independent existence of the electron within the nucleus became evident. For example, according to Heisenberg's Uncertainty Principle, for an electron to be confined to the nuclear volume its energy must necessarily be indeterminate to about 60 MeV, which is an unrealistically high value for the potential barrier preventing escape from the nucleus.

Rutherford suggested that the electrons did not have an independent existence in the nucleus, but that an electron and a proton were bound together to form a neutral particle which, while as yet hypothetical, was described as a *neutron*. Following experiments by Bothe and Becker (1930) and by Curie and Joliot (1932), which demonstrated the emission of highly penetrating radiation following the α-particle bombardment of beryllium and boron, Chadwick was, in 1932, able to demonstrate the existence of a particle with a mass very close to that of the proton and with no charge. Rutherford's neutron became a reality, enabling Heisenberg, in 1932, to propose a detailed proton-neutron theory of the nucleus, where the number of protons determined the nuclear charge Z, and in which $(A - Z)$ neutrons made up the remainder of the required nuclear mass.

The proton-neutron theory is in agreement with the evidence presented by radioactivity. The α-particle consists of two protons and two neutrons, and it and the β-particle can be formed at the time of emission, the former by simple combination and the latter by decay of a neutron into a proton, an electron, and another particle of essentially zero mass and charge known as the neutrino. Some theories hold that α-particles exist as such within the nucleus, but formation at the time of emission is considered to be the more probable.

Using the proton-neutron model of the nucleus it becomes possible to explain radioactive disintegration of heavy nuclei, to describe the composition of the nuclei of all the elements, to understand the transmutation of one element into another, and to predict the formation of heavier elements than those existing naturally, through transmutation processes.

A problem which is of great importance to our understanding of many phenomena related to the nucleus and its component nucleons, and which is still far from fully understood, is that of the forces which hold the nucleus together. The normal electrostatic repulsive force between two protons at separations less than a nuclear diameter is very large, and strong short-range attractive forces between nucleons must necessarily exist at these separations, in order to overcome the electrostatic repulsion. Outside the nucleus these attractive forces become insignificant and Coulomb repulsion governs the interaction between emitted nucleons, α-particles, and the parent nucleus. Similarly, the scattering of charged particles by the nucleus is a function of the electrostatic potential barrier unless the particles have themselves sufficient energy to penetrate the nucleus. We have therefore a picture of an internuclear potential which is repulsive, increasing in strength as the separation decreases to somewhere in the region of the nuclear radius, when it becomes an attractive potential well. The shape of the barrier at its peak, and the configuration of the attractive well itself, are both matters for conjecture, and assumptions concerning these are the first steps in establishing a number of proposed models for the nucleus.

Nuclear Models

We propose in this section to mention only briefly the basic postulates of several models for the nucleus, dwelling at greater length on the liquid drop model because of its relevance to the phenomenon of fission.

Theories of the motion of the nucleons inside a nucleus generally hinge on the type of field which they are assumed to experience, and whether or not we can consider each nucleon individually. For example, one model requires the existence of α-particles and their continual destruction and reformation through exchange of component nucleons. Another requires a statistical treatment of the particles in the nucleus. The shell model, a direct analogy to electron shells in atoms, arose from experimental observations of a periodicity of some nuclear properties over large ranges of proton and neutron content of the nucleus. Its success in this respect is perhaps strange in view of the basic premise, which requires each nucleon to move independently of all others in the average field of its interaction with the others. This independent motion is difficult to reconcile with the very strong interactions between individual nucleons. We would intuitively expect some interaction between nucleons and consequent collective motion. Such a collective nucleonic motion was the prime assumption of Bohr and Wheeler who, in 1939, compared the nucleus to a liquid drop. The nucleons were assumed to behave like molecules in a drop of liquid, being evaporated from the surface during radioactive disintegration of nuclei. Both the shell model and the liquid drop model have useful features in predicting many properties connected with nuclei, and it was therefore desirable to attempt some reconciliation between the two. This was achieved by Bohr (1951) and by Bohr and Mottelson (1953) with a *collective shell model*, which suggested that the entire shell configuration oscillates in shape, some nuclei, because they consist of closed shells, being less subject to this collective oscillation than others.

We have mentioned only briefly in this section the basic postulates of several proposed models for the nucleus. The liquid drop model will be considered further in the next section, because of its success in describing the phenomenon of fission.

The Discovery of Nuclear Fission

As seems to be often the case with major advances in our knowledge, the fission process manifested itself during the investigation of an unrelated phenomenon. During attempts to synthesize transuranium elements by the transmutation of uranium through neutron bombardment, it was observed that isotopes, at first wrongly estimated to be of radium, were present in the end-products of bombardment. The mistaken identification was rectified in convincing experiments by Hahn and Strassmann (1939), who showed beyond doubt that lighter elements resulting from neutron irradiation of uranium were isotopes of barium and lanthanum. It was difficult to explain the formation of these lighter atomic species unless some entirely new process was responsible, and Hahn and Strassmann were understandably extremely tentative in suggesting that, instead of the usual emission of a small particle or group of particles from the bombarded nucleus, the new transmuted uranium nucleus had become unstable and had split into two approximately similar fragments. Their suggestion, however, stimulated a repetition of the experiment in many laboratories, confirming their results and demonstrating the fission of uranium, thorium, and protactinium nuclei. The next few years saw an intense effort in the study of the fission process, although for almost a decade the military implications of fission interfered in the dissemination of much of the knowledge acquired.

The Fission Process

The naturally occurring isotopes of uranium are those with mass numbers 234, 235, and 238 or, in the common notation, including the relative percentage compositions:

$$_{92}U^{234} \qquad\qquad _{92}U^{235} \qquad\qquad _{92}U^{238}$$

$$0{\cdot}005\% \qquad\qquad 0{\cdot}71\% \qquad\qquad 99{\cdot}28\%$$

The $_{92}U^{235}$ isotope has a high cross-section for fission by thermal neutron capture, so that the reaction:

$$_{92}U^{235} + _{0}n^{1} \rightarrow [_{92}U^{236}] \rightarrow A + B \qquad (1.2)$$

in which A and B are the fission product nuclei, is highly favoured. The heavier isotope, $_{92}U^{238}$, requires much higher energy neutrons (~ 1 MeV) for fission.

When a uranium nucleus undergoes fission, the product nuclei, or *fission fragments*, separate with high energy, their momentum being equal and opposite. Almost always the nucleus splits into two fragments; although ternary fission has been observed to occur in $_{92}U^{235}$, the frequency being less than one for each 10^4 binary fission events. A cloud chamber photograph obtained by Bøggild, Brøstrom, and Lauritsen of two fragments from a binary fission event is shown in Figure 1. The total energy released during such an

Figure 1. Cloud chamber photograph of two fragments from a binary fission event. (After Bøggild, Brøstrom, and Lauritsen.)

event may be approximately estimated very simply by considering the Coulomb energy of two fragments at a separation equal to the sum of their nuclear radii. As we shall see later, the probability of obtaining two equal mass fragments from a fission event is small, but we do not significantly affect the calculated fission energy by assuming two equal fragments. If we do this, the Coulomb energy becomes:

$$E_{\text{fission}} = (Z_1 Z_2 e^2)/r$$

$$= \left\{ \frac{(46)^2 \times (48 \times 10^{-10})^2}{1 \cdot 5 \times 10^{-12}} \times \frac{1}{1 \cdot 6 \times 10^{-12}} \right\} \text{eV} \qquad (1.3)$$

$$\sim 200 \text{ MeV}$$

This figure may be arrived at independently by considering the energy equivalent of the mass discrepancy between the two sides of the equation for fission. Experimental confirmation of the figure came from several investigators, yielding primary energy releases from about 120 to 180 MeV.

In addition to the two fission fragments, the end products of fission include the so-called 'prompt' neutrons. The fission fragments immediately after the event are unstable, having an excess of neutrons, and they each emit one or two of these neutrons within a time of less than 10^{-15} sec. The emitted neutrons have thermal energies and can go on to cause fission in other uranium nuclei, initiating a chain reaction. The rate of progress of fission events can be controlled by absorption of the neutrons in non-fissile material – the principle of the nuclear reactor – or it can be allowed to 'run away', generating large amounts of energy very rapidly – hence its military application.

Bohr explained the fission process in two steps. Firstly, a compound nucleus was formed upon penetration of the neutron, and the excitation energy supplied by the neutron caused a type of thermal agitation in the nucleus. Secondly, the energy of this disturbance was transformed into potential energy of deformation, eventually leading to fission. In the next section we shall see how this distortion and splitting of the nucleus may be explained on the basis of the liquid drop model.

Fission on the Liquid Drop Model

The resemblance of the fission process to the breaking up of a drop of liquid following excitation was first pointed out by Meitner and Frisch (1939). They suggested that, in a similar fashion to a mechanically vibrating liquid drop, a nucleus excited into vibration by an absorbed neutron might develop a thin waist in the middle, and then the Coulomb repulsion of the two halves would overcome the surface tension of the nucleus, leading to the separation of two roughly similar fragments. The first thorough treatments of this process were made by Bohr and Wheeler and by Frenkel in 1939. They gave detailed descriptions of the fission of a uranium nucleus by fast and slow neutrons, and predicted successfully the spontaneous fission of heavier nuclei.

When the excitation energy for slow neutron capture causes the spherical nuclear 'drop' to oscillate, it first becomes ellipsoidal. If the surface tension forces tending to cause a return to the spherical shape are still more than counterbalanced by the distorting effect of the excitation, the ellipsoid tends to become further elongated and a waist forms. The nucleus is then of dumb-bell shape. Further pinching of the waist results in repulsion of the two halves followed by fission. The progress of this distortion and splitting is shown schematically in Figure 2. This of course is a very simplified picture, and the deformation of the spherical nucleus has more complicated modes than the first-order one illustrated here. The Bohr-Wheeler theory derives an equation for small deviations from the spherical form, assuming symmetrical distortions and a non-viscous nuclear fluid:

$$r = R[1 + a_0 + \sum_n a_n P_n(\cos \theta)] \tag{1.4}$$

where n and θ describe the distortions, P_n are Legendre polynomials, R is the non-distorted sphere radius, and a_n are the small constants which represent the distortion parameters.

By considering the surface tension of the distorted surface and the Coulomb energy of all pairs of charges within this surface, Bohr and Wheeler derived an expression for the total energy in the excited nucleus. Using this, they found

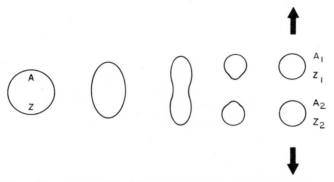

Figure 2. Schematic representation of the fission of a nucleus on the basis of the 'liquid drop' model.

that a spherical nucleus becomes unstable against small deformations and will not return to its spherical shape if:

$$Z^2/A \gtrsim 45 \tag{1.5}$$

Thus all nuclei with Z^2/A higher than this value were predicted to be unstable against spontaneous fission. We note that for $_{92}U^{235}$ the value of Z^2/A is 36; so that for this nuclide spontaneous fission is unlikely to occur.

Despite the undoubted success of the liquid drop model in explaining many aspects of the fission process, some objections are immediately evident. For example, firstly, the theory considers only small distortions of the excited nucleus, neglecting the possibility of larger asymmetrical disturbances. Secondly, a most serious defect in the theory is its failure to predict the experimental fact that the nucleus splits into two fragments of *unequal* mass and charge. In the liquid drop theory, the potential barrier is lowest for symmetrical splitting of the nucleus, and this persists even when large deformations are taken into account. Thirdly, the theory has been somewhat unsuccessful in predicting rates of spontaneous fission for the different nuclei.

A number of attempts were made through the years to modify the liquid drop theory in order to eliminate its defects. Fong (1956) developed a statistical theory of nuclear fission by employing an analogy with the theory of chemical equilibria, and this had some success in predicting that a pair of asymmetric fragments were energetically more favourable than an identical pair. The application of the nuclear shell model to fission was suggested by Meitner in 1950, and Hill and Wheeler in 1953 based their approach on the collective model, which considers the energy levels of the deformed nucleus. Some success was evident in explaining the angular distribution of the fission fragment emission, but the detailed explanation of the asymmetry of fission remains an open problem.

Spontaneous Fission

The spontaneous fission of uranium was first observed in 1940 by Petrzhak and Flerov, who noticed that one spontaneous fission occurred in 1 g of natural uranium approximately every 100 sec. The phenomenon was subsequently detected in several heavy nuclei ranging from thorium ($_{90}Th^{230}$) to curium ($_{96}Cm^{242}$), the fission rate in the transuranic nuclei being small compared to the rate of α-particle disintegration. To give some indication of the relative stability against spontaneous fission, as the atomic number increases, we might mention that the fission half life of $_{92}U^{235}$ is 2×10^{17} years, that of $_{92}U^{238}$ is 8×10^{15} years, for $_{94}Pu^{240}$ it is $1 \cdot 2 \times 10^{11}$ years, and for $_{96}Cm^{242}$ it is $7 \cdot 2 \times 10^{6}$ years. In general the logarithm of the half-life falls off as Z/A increases.

Fission of Lighter Nuclei

According to the liquid drop model the fission of any moderately complex nucleus should be a possibility, provided that a sufficiently large excitation is supplied. This has been experimentally verified by bombarding such materials as tantalum, platinum, and other heavy nuclei with high energy, artificially accelerated particles such as helium ions or deuterons. One puzzling result which has not been adequately explained is that the fragments from these nuclei do not show the strongly asymmetric mass distribution of the uranium fragments. By varying the energy of the fission-producing particle, however, it has been shown that the probability of symmetrical fission increases with increasing energy. The mechanisms involved in determining the relative masses of the two fission fragments is clearly quite complicated. Fission by proton bombardment of nuclei as small as copper has been observed. We shall only be considering the results of the fission of uranium nuclei.

The Products of the Fission Process

The stable fragments which remain after the thermal neutron induced fission of a U^{235} nucleus range in mass number from $A = 72$ to $A = 162$, although the large majority of fragments ($\sim 97\%$) fall into two groups, known as the 'light' and 'heavy' groups, centred respectively around $A = 96$ and $A = 137$. This is evident from the well-known twin-peaked fission yield curve of Figure 3. It is interesting to note that the numbers of neutrons in the most commonly produced fragments are those predicted for closed shells in the nuclear shell model. Shell effects have also been suggested as the cause of some fine structure observed in the light and heavy mass peaks.

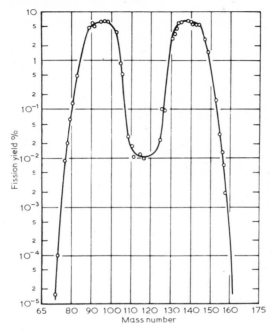

Figure 3. Fission yields of products with various mass numbers from uranium-235.

The immediate products of a fission event cannot possibly be stable, since the ratio of the number of neutrons to the number of protons in a uranium nucleus is about 1·6, whereas in the fission fragment range of atomic number the ratio is between 1·3 and 1·5. Hence there is a neutron excess in the fragments, and they decay to a more stable form by emission of β-particles and one or more 'delayed' neutrons. This process of β-emission represents an alternative method for the removal of excess neutrons, by changing them into protons, and a very common method for the decay of fission products into stable nuclides is through chains of β-decaying isobars. For example, the decay of xenon ($_{52}Xe^{143}$), a primary fission product, to the stable isotope of neodymium ($_{60}Nd^{143}$) takes place by the following chain:

$$_{54}Xe^{143} \rightarrow {}_{55}Cs^{143} \rightarrow {}_{56}Ba^{143} \rightarrow {}_{57}La^{143} \rightarrow {}_{58}Ce^{143} \rightarrow {}_{59}Pr^{143} \rightarrow {}_{60}Nd^{143}$$

$$(1.6)$$

This is an exceptionally long sequence; the majority of fission decay chains exhibit fewer steps.

The total energy of fission is shared among the fragments, the electrons, protons, neutrons, and neutrinos emitted by the fragments both at the time of fission and later (Table 1). Measurements of the kinetic energy of the fragments

TABLE I

Distribution of fission energy

	MeV
Kinetic energy of fission fragments	162
β-decay energy	5
γ-decay energy	5
Neutrino energy	11
Energy of fission neutrons	6
Instantaneous γ-ray energy	6
	195

show that, as in the case of their masses, there are two distinct groups. The energy spectrum of fragments from the fission of $_{92}U^{235}$ by slow neutrons is shown in Figure 4. Again, the asymmetry of the process is demonstrated by the peaks around 61 and 95 MeV. From momentum conservation during fission we can assign the low energy peak to the heavy fragment group and the high

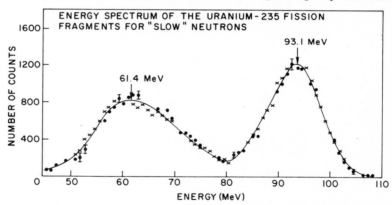

Figure 4. Energy spectrum of fragments from the fission of U^{235} by slow neutrons.

energy peak to the group of light fragments. Fission caused by fast neutrons up to 14 MeV gives two peaks situated in similar energy regions, although the minimum between the peaks is not so deep. For high energy neutrons of 90 MeV, the fission fragment energy spectrum contains only one peak. This is a further indication of the symmetry of fission at very high energies.

Measurement of the masses and kinetic energies of fission fragments requires relatively straightforward techniques. The energy is found by measuring the ionization produced in an ionization chamber, and assuming of course that the nucleus before fission is at rest, the masses follow from the conservation of momentum:

$$E_1/E_2 = (\tfrac{1}{2}M_1 V_1^2)/(\tfrac{1}{2}M_2 V_2^2) = M_2/M_1 \qquad (1.7)$$

since $M_1 V_1 = M_2 V_2$.

The charges carried by the fragments present a much more difficult problem since these will vary as the energetic particles travel through the medium. In addition to any nuclear charge imbalance caused by an initial excess of protons, the fragments carry along bound electrons. To the electrons of the uranium atom, fission of the nucleus obviously takes on catastrophic proportions, and they certainly do not divide themselves between the resulting fragments in proportion to the atomic numbers of the latter. Many, in fact, are left behind by the energetic fragments, which therefore begin their journey with a high charge. Approximate estimates of the initial charge use as a starting point the assumption that an electron will be bound to one or other fragment only if its average velocity in the uranium atom before fission is greater than the initial linear velocity v of one of the fragments. Thus the fragments carry along the K, L, and M shell electrons and some from the N shell, so that the initial positive charges are in the region of $18e^+$. Measurements of the ionic charge of fission fragments by deflection in a magnetic field give values of $16e^+$ for the light fragments and $14e^+$ for the heavy fragments. It should of course be stressed that the velocity dependence criterion is an extremely rough estimate of charge, and ideally one should endeavour to find a solution to the wave equation for a uranium atom during its separation into two fragments with a given relative velocity. Except in the limiting cases of zero relative velocity of separation, in which all electrons would be carried away bound to the fragments, and an extremely high velocity, in which both fragments would be completely stripped, the wave equation is too complicated to yield a solution.

Because of its high initial charge, the fission fragment ionizes atoms and molecules of the medium, in which it is moving, by Coulomb excitation. Electrons removed in this way may remain essentially free or they may be captured by the fragment, altering its residual charge. In addition, if the fragment approaches very close to an atom of the medium, a simultaneous catastrophic rearrangement of electrons on both fragment and atom may occur, the final electronic configuration being somewhat unpredictable. Eventually, of course, the fragment will slow down through energy loss to the medium and in doing so will capture electrons, finally becoming a neutral atom with only thermal velocity. The processes involved in the arresting of a fission fragment and in the attenuation of its energy as it passes through matter will be discussed later in greater detail.

The Problems of Anisotropy and Asymmetry

We have already mentioned the peculiar tendency of the nucleus under-going spontaneous or thermal neutron-induced fission to split into two asymmetric fragments, and have noted that symmetrical fission is unlikely unless the bombarding particle has high energy. To give a detailed account of the large volume of experimental and theoretical endeavour which has been devoted to an understanding of this asymmetry, so far with limited success, would require a book in itself. It is, nevertheless, fitting here to present a brief statement of the problem and the experimental results which define it, together with the current theoretical opinions as to the cause of the asymmetry. Before we do this, however, we must mention experimental observation of a phenomenon which is very probably strongly interlinked with that of the inequality of fragment mass, namely the angular anisotropy in the direction of fragment emission.

It has been shown that in a number of cases there is a correlation between the directions of emission of fragments and the direction of approach of the particle producing fission. This was first observed in experiments on the fission of thorium through irradiation with γ-quanta, when the fission fragments emerged mainly perpendicular to the γ-ray beam. A significant fact was that this angular anisotropy closely followed the mass asymmetry. Little angular variation was noticed in cases where fission was symmetric and, as in the case of the mass asymmetry of fission, the angular anisotropy was found to decrease at high particle energies. When the fission was caused by particles such as neutrons or protons, the predominant directions of emission were parallel to the momentum vector of the bombarding particles – at $0°$ or $180°$ – with minima at $90°$ and $270°$.

A theory which attempted to explain the angular anisotropy was proposed by Hill and Wheeler in 1953. The basis of the theory was that the nuclear 'drop' was not spherical before fission, and therefore the deformation which led to fission would be most likely to orient itself along the symmetry axis. It was further necessary to assume that the probability that a particle will cause fission in a nucleus depends on the direction of the beam of particles relative to the spheroidal symmetry axis of the nucleus. Fission must occur with highest probability when the axis is perpendicular to the direction of the momentum vector given to the nucleus by the particle. Thus, for photofission, the maximum number of fragments would be emitted perpendicular to the beam, whereas for nucleon-induced fission more would be emitted parallel to the beam. This is in agrement with experiment.

While it appears from experimental evidence that the angular anisotropy and mass asymmetry are related, it is not feasible to draw the most obvious conclusion from this fact, namely that fission may be divided into symmetric and asymmetric fission processes, where isotropic fission is associated with symmetric fission, and anisotropic with asymmetric. Anisotropy also appears in fission caused by fast particles, where symmetric fission is predominant. It is evident, therefore, that there is no sharp division, the two types coexisting in different proportions under different conditions.

Most attempts to explain the angular anisotropy and mass asymmetry have been essentially of a qualitative nature, and, because of their fundamental assumptions regarding the model of the nucleus undergoing fission, their success has been limited. The problem is extremely complicated and, while the fission of a heavy nucleus is fairly well understood in broad outline, the detailed results impress us with the fact that there is much we can only explain by hypothesis, owing to the inadequacy of our present models for the nucleus. It is possible that, in addition to giving us more information on the phenomenon of fission, further experimental and theoretical work on the process may deepen our insight into the atomic nucleus itself, its equilibrium configuration, and the collective interactions of its component nucleons.

Conclusion

We have in this chapter endeavoured to present an account of the phenomenon of nuclear fission, showing how events led from the discovery of radioactivity to the first detection of the sundered atom, discussing briefly the important aspects of the process and their implications in connection with our concept of the nucleus. Of necessity the story has been brief. For more complete accounts we would refer the reader to some of the many items and reviews which have appeared in the literature since 1939. Our present purpose has been to introduce the fission fragment and to discuss the violent circumstances of its conception and birth. In what follows we shall consider the arresting of this heavy energetic particle by the solid state.

References

BECQUEREL, H. (1896), *Compt. Rend.* **122**, 420.

BØGGILD, J. K., BRØSTROM, K. J., and LAURITSEN, T. (1940), *Phys. Rev.* **58**, 651.

BOHR, A. (1951), *Phys. Rev.* **81**, 134.

BOHR, A., and MOTTELSON, B. R. (1953), *Kgl. Danske Vidensk. Selskab. Mat.-fys.* **27**, No. 16.

BOHR, N. (1936), *Naturwiss.* **24**, 241.

BOHR, N., and WHEELER, J. (1939), *Phys, Rev.* **56**, 426.

BOTHE, W., and BECKER, H. (1930), *Z. Physik* **66**, 289.

CHADWICK, J. (1932), *Nature* **129**, 312.

CURIE, I., and JOLIOT, F. (1932), *Compt. Rend.* **194**, 273 and 708.

FONG, P. (1956), *Phys. Rev.* **102**, 434.

FRENKEL, I. (1939), *J. Exp. Theor. Phys.* **9**, 641.

GAMOW, G. (1928), *Z. Physik* **51**, 204.

HAHN, O., and STRASSMANN, F. (1939), *Naturwiss.* **27**, 89.

HEISENBERG, W. (1932), *Z. Physik* **77**, 1.

HILL, D. L., and WHEELER, J. A. (1953), *Phys. Rev.* **89**, 1102.

MEITNER, L. (1950), *Nature* **165**, 561.

MEITNER, L., and FRISCH, O. R. (1939), *Nature* **143**, 239.
PETRZHAK, K. A., and FLEROV, G. N. (1940), *J. Exp. Theor. Phys.* **10**, 1013.
RUTHERFORD, E. (1911), *Phil. Mag.* **21**, 669.
RUTHERFORD, E., and SODDY, F. (1903), *Phil. Mag.* **5**, 576.

Reviews

HALPERN, I. (1959), *Ann. Rev. Nuc. Sci.* **9**, 245.
TURNER, L. A. (1940), *Rev. Mod. Phys.* **12**, 1.
WALTON, G. N. (1961), *Quart. Rev. Chem. Soc.*, London, **15**, No. 1.
WHITEHOUSE, W. J. (1952), *Prog. Nucl. Phys.* **2**, 120.
Proceedings of the United Nations International Conferences on the Peaceful Uses of Atomic Energy, Geneva (1955 and 1958 – obtainable from the United Nations).
Proceedings of the Symposium on the Physics of Fission, Chalk River, Ontario, Report AECL-329 (1956 – obtainable from Atomic Energy of Canada Ltd.).
Supplement No. 1 of *Atomnaya Energiya*, translation 'Physics of Nuclear Fission', Pergamon Press, London (1958).

CHAPTER TWO

Passage of Fission Fragments through Crystalline Matter

The way in which a fission fragment is born through the splitting of a heavy nucleus with the liberation of a large amount of energy has already been described. We now consider further the situation from the point where two fragments of a nucleus which has just undergone fission are separating with kinetic energies of up to 100 MeV. Unless the parent nucleus is completely isolated in a perfect vacuum, these fragments will immediately interact with atomic and nuclear particles on or close to their path. For example, if the fission process takes place in a uranium foil undergoing thermal neutron bombardment, then the resulting nuclear fragments will lose a significant portion of their energy of fission to that part of the uranium foil which lies on their escape route to the surface. It may be shown that this effect is sufficient to convert the two characteristic energy peaks which were described in Chapter 1 to a single peak at lower energies (~ 30 MeV). In the present chapter, therefore, it is our purpose to provide qualitative answers to the following questions:

(1) By what processes does a fission fragment lose energy to the constituents of the material in which it is travelling?

(2) To what extent do these processes depend on the energy and effective charge of the fragment?

(3) How is the energy lost by the fragment dissipated in the environment of its trajectory in the material?

(4) What, in general terms, is the form of the permanent damage in a solid through which a fission fragment has passed?

(5) What is the eventual fate of a fission fragment brought to rest in a solid?

The Energy Loss Processes

If we ignore the relatively unlikely mechanism of nuclear transmutation, a heavy charged particle moving through a medium composed of atomic nuclei and electrons can lose energy by three main processes. At the energies which fission fragments initially possess (10^7 eV), the principal mode of energy loss is that due to interaction with the atomic *electrons* by Coulomb excitation or ionization. A fragment may also lose energy by direct collisions with *nuclei* of the stopping medium, though such events are rare until the particle nears the

15

end of its path. The third energy loss process, that of emission of 'brems-strahlung' and Cerenkov *radiation*, is small compared to either of the first two processes. For any energy, E, therefore, the total rate of energy loss (the absolute stopping power) is a combination of the rates of energy loss due to each of the three processes, and may be written in the form:

$$\left(-\frac{dE}{dx}\right)_{total} = \left(-\frac{dE}{dx}\right)_{electronic} + \left(-\frac{dE}{dx}\right)_{nuclear} + \left(-\frac{dE}{dx}\right)_{radiation}$$

The problem is then decoupled, and each of the three contributions may be examined separately.

Energy Losses by Radiation

Because of its relative unimportance in contributing to the total energy loss, we shall mention this process only briefly. 'Bremsstrahlung', or braking radia-tion, is emitted when a charged particle is rapidly accelerated or retarded in the field of a nucleus. The intensity of the radiation is, according to classical electrodynamics, inversely proportional to the square of the mass of the incident particle, and consequently the phenomenon is of much less importance in the case of heavy ions than it would be for lighter particles such as electrons. Cerenkov radiation arises from the polarization of a crystal along the path of a charged particle when the velocity of the latter exceeds the phase velocity of light in the medium. This polarization is time-dependent owing to the moving electric field of the particle, and can produce a radiation field. The effect is, of course, vanishingly small in the case of a fission fragment, since the particle velocity is very much less than the velocity of light in any crystal.

Nuclear Collision Losses

Retardation of a fission fragment by collisions with atoms in the crystal only takes precedence near the end of the fragment range, where electronic excita-tion becomes small. Such two-body 'nuclear' collisions may be described classically (Bohr, 1948) provided that the collision diameter b, the distance of nearest approach in a head-on collision, is very much greater than the de Broglie wavelength λ. For scattering of a charged particle by a Coulomb field of force (Chadderton, 1965), this condition may be written:

$$\kappa = b/\lambda = (Z_1 Z_2 e^2)/(hv) \gg 1 \qquad (2.1)$$

where Z_1 and Z_2 are the atomic numbers of the moving and struck particles respectively and v is the velocity.

The most simple way classically to describe a 'nuclear' collision between two heavy ions does, in fact, employ the Rutherford scattering law. The differential cross-section for scattering through an angle θ into a solid angular increment $d\omega$ is given by:

$$d\sigma_{\text{nuclear}} = \left(\frac{Z_1 Z_2 e^2}{2M_0 v^2}\right)^2 \text{cosec}^4\left(\frac{\theta}{2}\right) d\omega \tag{2.2}$$

where M_0 is the reduced mass. Alternatively, in terms of energy transfer T, for collision between two particles specified by (M_1, Z_1, E) and $(M_2, Z_2, 0)$ given by:

$$d\sigma(E, T)_{\text{nuclear}} = \frac{\pi Z_1^2 Z_2^2 e^4}{E}\left(\frac{M_1}{M_2}\right)\frac{dT}{T^2} \tag{2.3}$$

This description, however, assumes that the collision is occurring between two point charges and masses with central fields of force. In the case of a collision between a fission fragment and an atom, the picture is complicated by the presence of bound electrons on each nucleus, and the problem is essentially a many-body collision process. It is possible to approximate a real situation by considering the *screening* effect of the bound electrons, replacing the simple Coulomb potential of Rutherford scattering by:

$$V(r) = (Z_1 Z_2 e^2/r)\exp(-r/a) \tag{2.4}$$

where a is a *screening parameter*.

In these circumstances the conditions under which we must replace the Rutherford scattering law by one which allows for screening are determined by the value of the ratio:

$$\zeta = b/a \tag{2.5}$$

Thus, only if the collision diameter is small compared to the screening parameter ($\zeta \ll 1$) will the Rutherford Law be valid (weak screening). This corresponds to high fragment velocities.

We note, for completeness, that the condition for the validity of a classical approximation to two-body collisions in crystals is modified by the introduction of the general concept of screening. To the requirement that $\lambda \ll b$ must be added the condition that $\lambda \ll a$, so that:

$$\kappa \gg \xi \tag{2.6}$$

This condition is evidently satisfied for the fission fragment, whatever the velocity. The classical approximation is therefore also sufficient for strong screening, when $\xi \gg 1$. In this regime the screening parameter a is small, so that collisions may effectively be considered as occurring between rigid spheres. Giving the hard spheres radii, $R(E)$, which depend on the energy, the differential scattering cross-section is then given by the formula:

$$d\sigma(E, T)_{\text{nuclear}} = 4\pi[R(E)]^2 (dT/T_{\text{max}}) \tag{2.7}$$

where:

$$T_{\text{max}} = \frac{4M_1 M_2}{(M_1 + M_2)^2} \cdot E \tag{2.8}$$

In the case of weak screening (Rutherford scattering), therefore, the differential cross-section depends on the inverse square of the energy transfer T; so, at high fragment velocities, low energy transfers are preferred. For strong screening (hard sphere scattering), which is clearly the situation at low fragment velocities, all energy transfers up to T_{max} are equally probable.

The differential cross-section $d\sigma(E, T)$ is very simply related to the specific rate of energy loss of the fission fragment by the following:

$$\left(-\frac{dE}{dx}\right)_{\text{nuclear}} = n_0 \int_{T_{min}}^{T_{max}} T \, d\sigma(E, T) \tag{2.9}$$

where n_0 is the spatial density of available nuclei and T_{min} is a lower limit of the order of $0{\cdot}1$ eV. We may illustrate the determination of T_{min} for the case of weak screening. By a simple geometric treatment of the scattering process, since the deflection angle θ_{min} for transfer of energy T_{min} is small, it may be shown that (Bohr, 1948):

$$\theta_{min} = b/p_{min} \tag{2.10}$$

and

$$T_{max}/T_{min} = (2/\theta_{min})^2 \tag{2.11}$$

where p_{min} is the corresponding impact parameter. Now the Rutherford scattering law will clearly only hold for values of θ for which p is small in comparison with the screening radius a. Taking this as the limit, therefore:

$$\theta_{min} = \xi \tag{2.12}$$

so that:

$$T_{max}/T_{min} = (2/\xi)^2 \tag{2.13}$$

where:

$$T_{min} = \frac{Z_1^2 Z_2^2 e^4}{a^2} \left(\frac{M_1}{M_2}\right) \frac{1}{E} \tag{2.14}$$

The specific rate of nuclear energy losses then becomes:

$$\left(-\frac{dE}{dx}\right)_{\text{nuclear}} = \left(\frac{2\pi Z_1^2 Z_2^2 e^4}{M_2 v^2}\right) n_0 \log_e\left(\frac{T_{max}}{T_{min}}\right)$$

$$= \left(\frac{\pi Z_1^2 Z_2^2 e^4}{E}\right)\left(\frac{M_1}{M_2}\right) n_0 \log_e\left[\frac{2aE}{Z_1 Z_2 e^2}\left(\frac{M_2}{M_1 + M_2}\right)\right] \tag{2.15}$$

Electronic Energy Losses

If it is also assumed, in a first approximation, that energy loss collisions which the moving fragment makes with atomic electrons are also to be considered within the framework of classical mechanics, then the differential scattering cross-section is again given by the Rutherford law:

$$d\sigma(E, T)_{\text{electronic}} = \frac{\pi Z_1^2 e^4}{E} \left(\frac{M_1}{m_e}\right) \frac{dT}{T^2} \tag{2.16}$$

where m_e is the electronic mass. Choosing I, the ionization energy of the electron from the atom, as a suitable lower limit $T_{\min}$, and defining the maximum energy transfer $T_{\max}$ in the normal manner:

$$T_{\max} = \frac{4M_1 m_e}{(M_1 + m_e)^2} \cdot E \simeq \frac{4m_e}{M_1} \cdot E \tag{2.17}$$

the appropriate classical equation for the specific electronic energy loss rate becomes:

$$\left(-\frac{dE}{dx}\right)_{\text{electronic}} = \left(\frac{\pi Z_1^2 e^4}{E}\right)\left(\frac{M_1}{m_e}\right) n_0 \sum_i \log_e \left(\frac{4m_e}{M_1} \cdot \frac{E}{I_i}\right) \tag{2.18}$$

where the summation extends over the i electrons in each of n_0 atoms, using appropriate values of the ionization potential I_i. This formula is, of course, strictly approximate. In the general case of passage of a charged particle through a crystal it is necessary to differentiate between 'resonance' and 'free' collision contributions to the electronic stopping power, the former being quantum mechanical in every way, and the latter classical. The way in which Bohr (1948) takes care of this in terms of the old quantum theory is referred to in Chapter 12.

Bethe (1930) and Bloch (1933) have also derived appropriate expressions for the electronic energy loss rate. Based on the Born approximation, Bethe's formula is the following:

$$\left(-\frac{dE}{dx}\right)_{\text{electronic}} = \left(\frac{2\pi Z_1^2 e^4}{E}\right)\left(\frac{M_1}{m_e}\right) Z_2 n_0 \log_e \left(\frac{4m_e}{M_1} \cdot \frac{E}{\bar{I}}\right) \tag{2.19}$$

where $\bar{I}$ is a suitable average taken over all excitable electrons, which may be approximated thus:

$$\bar{I} = kZ_2 \, (k \simeq 11 \cdot 5 \text{ eV}) \tag{2.20}$$

It is to be noted that, in the Bethe formula, the effect of including the small energy transfer processes – the 'resonance' processes of Bohr – is merely to double the rate of energy loss whilst retaining the same dependence on the fundamental physical parameters associated with stopping. The Bloch formula, however, contains the quantum-theoretical formula of Bethe, and an early classical formula of Bohr (1913), as limiting cases for small and large values of the parameter $\kappa = (Z_1 e^2)/(\hbar v)$. This equation has the following form:

$$\left(-\frac{dE}{dx}\right)_{\text{electronic}} = \left(\frac{2\pi Z_1^2 e^4}{E}\right)\left(\frac{M_1}{m_e}\right) Z_2 n_0 \left\{ \log_e \left(\frac{2m_e v^2}{\bar{I}}\right) + \psi(1) - \mathscr{R}\,\psi(1 + i\kappa) \right\}$$

$$\tag{2.21}$$

where ψ is the logarithmic derivative of the Γ-function and $\mathscr{R}\psi$ denotes the real part of ψ. For $\kappa = 0$, it is the Bethe formula (now familiarly referred to as the Bethe-Bloch formula), and for $\kappa \gg 1$, since $\psi(1) = -0.577$ and $\mathscr{R}\psi(1 + i\kappa) \simeq \log_e \kappa$, it is the Bohr formula:

$$\left(-\frac{dE}{dx}\right)_{\text{electronic}} = \left(\frac{2\pi Z_1^2 e^4}{E}\right)\left(\frac{M_1}{m_e}\right) Z_2 n_0 \log_e \left(\frac{1 \cdot 123 m_e v^2}{\kappa \bar{I}}\right) \qquad (2.22)$$

where $\omega = \bar{I}/\hbar$ is an average oscillation frequency for the electrons in the atom.

It should be emphasized here that the difficulties associated with a proper accounting for all of the possible eventualities in high energy collisions with an atom are profound. Thus the quantum mechanical treatment assumes that the collision is over before the electrons respond to the field of the moving fragment, which cannot be the case. Furthermore, the charge on the fragment fluctuates and falls as it penetrates into the crystal. To some extent it is possible to allow for the fall of the charge on the fragment by assuming that the particle only carries along those electrons whose orbital velocities are greater than the fragment velocity itself. Utilizing this hypothesis and the Thomas-Fermi model of the atom, the effective fragment charge becomes:

$$Z_1^{\text{eff}} = Z_1^{1/3} \hbar v/e^2$$

Likewise, the number of electrons of a target atom which may be removed by a fragment of velocity v is given by:

$$n = Z_2^{1/3} \hbar v/e^2 \qquad (2.23)$$

so that the electronic stopping power may be apportioned in some crude fashion to the various electronic shells. By proceeding in this fashion the energy loss rate to electrons assumes the following form:

$$\left(-\frac{dE}{dx}\right)_{\text{electronic}} = \frac{2 Z_1^{2/3} Z_2^{1/3} h^3 n_0}{\pi^2 e^2 m_e} \left(\frac{2E}{M_1}\right)^{1/2} \qquad (2.24)$$

For a fuller treatment the reader is referred to the original work of Bohr (1948).

The Absolute Stopping Power

The total rate of energy loss of a fission fragment passing through a crystal may, neglecting radiation losses, be expressed by a summation of two terms derived from analyses similar or identical to those presented above. The respective contributions of the two components at different fragment energies is best illustrated by an example. Figure 5 therefore shows both the total stopping power in ergs per centimetre and the electronic and nuclear components for the case of a median-light fission fragment from uranium travelling in a crystal of lead iodide (see Chapter 9). Weighted mean values of $M_2(96)$ and $Z_2(39)$ were employed in the calculation. These curves demonstrate the relative insignificance of nuclear collision energy losses until the fragment nears the end of its range, though it should be noted that nuclear collisions become more significant at higher energies if we employ the experimental fragment charge. A further underestimation of the nuclear collision contribution arises from use of the screened Coulomb interatomic potential, which falls off much more rapidly with distance than does the true potential. The contribution of a large number of glancing collisions to $(-dE/dx)_{nuclear}$ is lost.

Graphical integration of the curve for $(-dE/dx)_{total}$ yields a range-energy curve for the fragment. Figure 6 shows curves for both median-light and median-heavy ($M_1 = 137$, $Z_1 = 55$) fragments in lead iodide.

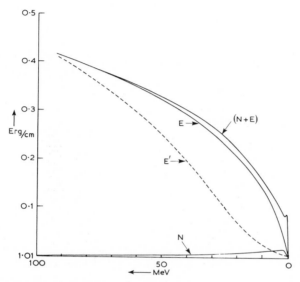

Figure 5. Graphical representation of the stopping power (N + E) for a median-light fission fragment travelling in a crystal of lead iodide. Approximations due to Bohr have been used in the calculations and the treatment permits an estimate of the relative contributions due to nuclear impacts (N) and electronic excitation (E). The broken line (E′) indicates that a reduction of the contribution due to electronic excitation is to be expected if experimental values for the variation of fragment charge are employed in the calculation.

Damaging Mechanisms

Having investigated the processes of energy transfer from the fission fragment to a stopping-medium, it becomes appropriate briefly to consider the general ways in which this energy is dissipated in a crystal, and the broad types of damage produced. We have divided collision events initiated by the energetic fragment into electronic and nuclear, into inelastic and elastic collisions. Let us now examine the processes which take place in the solid subsequent to each type of collision in order to establish which effects might be likely to be observable on a macroscopic scale. It will be fitting to assume a 'general' solid, although of course fission fragments produce quite different types of damage in the differing classes of materials. Such variations, however, will become apparent in later chapters, where different irradiated specimens will be considered separately.

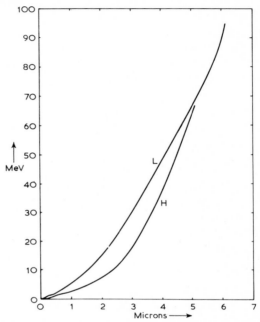

Figure 6. Range/energy relations for median-light (L) and median-heavy (H) fragments travelling in a crystal of lead iodide.

Atomic Displacement and Displacement Cascades

Although a major conclusion of the analyses of energy loss was that direct collisions with nuclei are of minor importance at high energies, it must not be assumed that atomic displacement is negligible. There is high probability that an atom as a complete entity will receive kinetic energy in that region of high thermal and electronic excitation which lies in the wake of a fission fragment. Therefore a number of atoms will be displaced by secondary processes arising from the electronic part of the energy loss of the fragment. Secondary displacement processes are also important in the history of a primary knocked-on atom deriving from a nuclear collision. Such an atom will undergo elastic and, if the energy is sufficient, inelastic collisions with its neighbours, initiating further secondary and higher order displacements. A cascade of displaced atoms will build up until the maximum energy of any atom falls below the appropriate *displacement threshold* E_d. The value of E_d depends on the binding and crystallographic characteristics of the material, and is normally about 25 eV, though as we shall see in later chapters it is directional-dependent in crystalline matter, and can be much higher in certain unfavourable directions.

A number of theories of the displacement cascade have been established in order to explain both the dissipation of energy in the material and the resultant permanent effects (Felder, 1967). In the very simple case where an atom in an element receives energy in excess of its displacement threshold, but insufficient for it to displace another, we can postulate with reasonable accuracy that it will eventually either return to its lattice site or come to rest in an interstitial position of metastable equilibrium, creating a Frenkel pair (assuming of course that replacement does not occur). When the energies involved are much higher, and a large cascade is probable, statistical treatment of the energy distribution among initial and succeeding displacements gives integral equations for the number of displaced atoms in the cascade. A theory intended to produce this number, however, inevitably begins with several assumptions which lead to over-simplification:

(1) The displacement threshold E_d is sharp and invariant. This neglects any directional or thermal effects.

(2) Collisions are purely elastic from an energy E_d up to an energy sufficient to ionize an atom. No displacements occur at energies above the latter.

(3) Two-body interactions only are assumed. No allowance is made for many-body processes or interaction between displaced atoms in a damaged region.

(4) No energy losses less than E_d are considered.

(5) Spontaneous annealing of damage through interstitial-vacancy annihilation is assumed negligible. Such processes reduce the number of permanent displacements and are especially significant in a material such as an ordered alloy, where two atomic species are present, replacement of one by the other leading to permanent disordering of the lattice.

On the basis of these assumptions we may nevertheless derive by a very simple model an expression for $\nu(E)$, the number of displaced atoms resulting from a primary of energy E. No further fundamental inaccuracy is introduced by assuming hard-sphere scattering, in which the two participating atoms have equal probability of possessing any final energy up to E. Thus, as a further simplification, let the initial energy in each collision be equally shared between the two participants. Then at any point in the progress of the cascade the energy is shared equally by all the atoms displaced at that stage. No further displacement occurs when the energy drops below $2E_d$, so we may immediately write for $\nu(E)$ the expressions:

$$\nu(E) = \begin{cases} E/(2E_d) & E > 2E_d \\ 1 & E_d < E < 2E_d \\ 0 & E < E_d \end{cases} \tag{2.25}$$

We remember that, in the energy regime above the lower limit for ionization E_I, no further displacements are allowed, so that:

$$\nu(E) = E_I/(2E_d) \qquad E > E_I \tag{2.26}$$

A more rigorous account, taking note of the energy required to overcome the potential well in order to displace an atom, leads to the asymptotic form:

$$\nu(E) = 0.36(E/E_d) \qquad (2.27)$$

It is hardly surprising, in view of the simplified model, that, when calculated values of $\nu(E)$ are compared with the results of carefully conducted experiments, the agreement is poor. Theory over-estimates the damage, in some cases by a factor of 10. Some of the blame for this lies with the assumptions listed above, but there are other factors associated with the concentration of damage into a small atomic volume and with the overlap of damage regions. The displacement cascade is, in reality, a very complicated, many-body process which we have at present no true way of describing, either analytically or by computational means.

The Spike Concept

Instead of considering the participation of just two atoms in any one collision event, now suppose that a passing fission fragment in a glancing collision gives an atom of the medium an energy less than E_d, so that it begins to vibrate in its potential well. We can easily picture the vibration spreading to its neighbours, developing a localized excitation which is effectively the same as if that small region of the solid were raised to a high temperature. The vibrations will decrease rapidly in amplitude as the excitation spreads from its source until thermal equilibrium is regained. Such localized excitations are termed *thermal spikes*. Although the temperature concept does not rigorously apply, since equilibrium never exists within the spike, it is possible to obtain useful information regarding the dimensions and excitation profile of the spike by application of the macroscopic laws of heat conduction. Thus Dienes and Vineyard (1957) calculate that a spherical thermal spike of 300 eV initial energy in copper will raise the temperature of a region of diameter 30 Å to the melting-point (1086°C) in a time of 5×10^{-12} sec, and that the cooling-period will be of the order of 10^{-11} sec. Because of the short duration, the spike region does not attain a true liquid state, and may be more accurately described as a superheated solid.

An energetic fission fragment passing through a crystal will certainly produce a great deal of thermal spike activity along its path, both by direct interaction and indirectly through the primary and secondary knock-ons which it generates. In this case the heat conduction equations are solved using cylindrical geometry, assuming a continuous spike along the fragment trajectory (Chapter 12). Such calculations indicate that a cylindrical volume of radius 100 Å and several microns in length may attain temperatures of the order of 10^4 °C.

In the case of the fission fragment this situation is, of course, further complicated by the large number of atomic displacements and displacement cascades, which in turn produce more thermal excitation. In particular, many neighbouring atomic displacements will be created in many small volumes adjacent to the trajectory of the fragment. This multiple displacement idea leads naturally to the concept of the *displacement spike*, first suggested and developed by Brinkman (1954, 1959).

The calculations which led Brinkman to present a model for the displacement spike involved estimations of the mean-free-path for the production of energetic knock-ons on the path of a fission fragment travelling through a

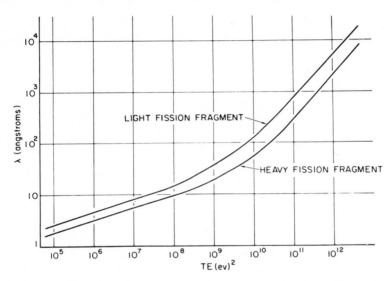

Figure 7. Energy dependence of mean-free-path (λ) of median light and median heavy fission fragments in uranium between production of successive knock-ons having energies in excess of *T*. *E* is here the energy of the fragment.

solid. Using the 'impulse approximation', since the fragment energy E is large compared to the energy transfer T in any single collision, and a suitable empirical interatomic potential, it is possible to derive an expression for the energy transfer T as a function of E and the impact parameter p. The mean-free-path λ is also a function of p, the link being the cross-section $\sigma(E, T^+)$ for transfer of energy *greater than T*:

$$\sigma(E, T^+) = \pi p^2 \qquad (2.28)$$

$$\lambda(E, T^+) = 1/[n_0\, \sigma(E, T^+)] \qquad (2.29)$$

where n_0 is the atomic density.

Curves showing λ as a function of the product TE for median-light and median-heavy fission fragments in uranium are presented in Figure 7. By selecting any fission fragment energy, the mean-free-path for energy transfer greater than any chosen value may be immediately obtained.

Such a calculation may be performed for any particle moving in any material using a suitable interatomic potential. For copper, if T is 25 eV (assumed to be the displacement threshold), a primary knocked-on atom moving through the lattice with about 1000 eV transfers energy greater than T with a mean-free-path which is *less than one interatomic distance*. This means that when the energy of a primary knock-on drops to this order of magnitude it quickly generates a multiple displacement of neighbouring atoms. Brinkman hypothesized that a multiple vacancy would therefore be created at the end of the path of an incident fission fragment or an energetic primary knock-on, the configuration immediately following each event being a shell of interstitial atoms surrounding a vacancy core (Figure 8). It is doubtful whether such a configuration is

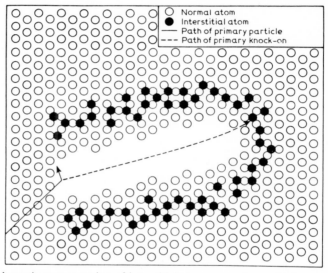

Figure 8. Schematic representation of interstitial atoms around a multiple vacancy during production of a Brinkman displacement spike.

stable, however, since the high density of interstitials exerts a high pressure on the central region. Most likely some collapse of the spike will occur, with interstitial-vacancy annealing, but in the final configuration dislocation loops and small regions of misorientation are highly probable. The effect is expected to be even more drastic in a compound or alloy, where disordering will greatly hinder the annealing process and increase the strains present in the final lattice configuration.

An additional important factor is that of the ordered lattice structure in a crystal. The phenomena of 'focusing' and 'channelling' in crystal lattices will be discussed later, but the fact that these directional effects enhance the long-range transport of both energy and matter in the lattice should not be omitted. Seeger (1962) therefore proposed a modification of the displacement spike which included the propagation of focused collision sequences along close-packed rows of lattice atoms. These sequences present a very good mechanism for the far separation of an interstitial at the end of the sequence from the corresponding vacancy at the beginning. The result is a high local density of vacancies in the region where the primary energetic particle is brought to rest, surrounded by only a few nearby interstitials; the so-called *depleted zone* (Figure 9).

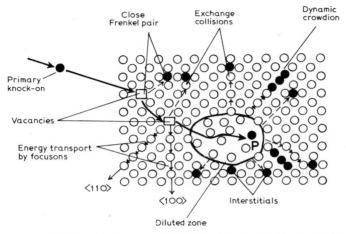

Figure 9. Schematic two-dimensional representation of the Seeger depleted zone and its environment in a face-centred cubic crystal. A primary knock-on impinges from the left and comes to rest on the site P.

The Fission Fragment Track

We are now in a position empirically to describe the damage produced by a fission fragment, and to offer a '*fission spike*' model for its track in a crystalline solid. There is a readily available estimate of the variation of the effective fragment charge with energy which yields the energy communicated to the electrons and hence the radius of the cylindrical thermal spike. Since the effective charge decreases as the fragment energy falls, this radius will also decrease, and to an approximation we may represent the thermal spike as a right circular cone with its apex at the final resting place of the fragment (Figure 10(a)). To include the effect of atomic displacements it will be recalled that collisions with atoms become more frequent as the fission particle slows down, so that the frequency of displacement cascade production will increase as more primary knock-ons come to rest, each creating spikes according to the schematic representation of Figure 10(b). The final picture of the fission spike will therefore be a superposition of Figures 10(a) and (c), the displacement spike production increasing in frequency until the end of the track, which itself takes the form of one large spike caused by the fragment as it comes to rest as an *impurity* atom in the lattice.

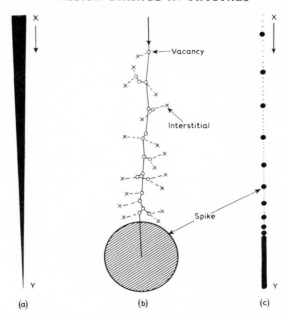

(a)   (b)   (c)

Figure 10. Schematic representation of the fission spike. (a) The fragment enters the crystal at X and comes to rest at the point Y. The intensity of the thermal spike at any point along this 'track' is given by the radius of a right circular cone (seen here in section) at that point. (b) Diagram showing the formation of a displacement spike by an energetic knock-on deriving from a collision between the fragment and a lattice atom. When the mean-free-path for atomic displacement falls below an interatomic distance a displacement spike is formed. (c) The frequency of displacement spike formation increases as the fragment slows down and there is a gradual transition from 'Rutherford' to 'hard sphere' collisions. The statistical nature of the frequency of collision events is considered to be understood and is omitted from the diagram.

References

BETHE, H. A. (1930), *Ann. d. Phys.* **5**, 325.

BLOCH, F. (1933), *Ann. d. Phys.* **16**, 285.

BOHR, N. (1913), *Phil. Mag.* **25**, 10.

BOHR, N. (1948), *Mat. Fys. Medd. Dan. Vid. Selsk.* **18**, 8.

BRINKMAN, J. A. (1954), *J. Appl. Phys.* **25**, 96.

BRINKMAN, J. A. (1959), *Nuclear Met.* **VI**, 1.

CHADDERTON, L. T. (1965), *Radiation Damage in Crystals*, Methuen (London).

DIENES, G. J., and VINEYARD, G. H. (1957), *Radiation Effects in Solids*, Interscience (New York).

FELDER, R. M. (1967), *J. Phys. Chem. Sol.* **28**, 1383.

SEEGER, A. (1962), *Radiation Damage in Solids*, IAEA (Vienna), **I**, 101.

CHAPTER THREE

Experimental Methods of Irradiation

We now pass to a brief consideration of experimental aspects of the study of fission fragment damage in solids, and in this chapter describe the methods which may be used to irradiate the specimens under investigation with fission fragments, prior to their observation by electron or field-ion microscopy. The preparation of particular classes of crystalline specimens suitable for study by these methods is presented in those chapters which deal with damage in different materials.

Techniques of Fission Fragment Irradiation

The most simple method for bombardment of a specimen with fission fragments is to expose it to the background of radiation in a nuclear reactor. However, the length of time required to produce appreciable damage in a specimen a few millimetres across is prohibitive, since the fission flux throughout the reactor is comparatively small. When we add to this the fact that in many experiments we wish to produce particle tracks in a specimen and therefore require the fission fragments to travel through at a low angle of incidence, it is easily seen that the direct exposure to reactor background fission fragments is both impractical and uneconomical as a method for the production of damage.

The most obvious solution to the problem of insufficient flux is to localize the source of fission fragments by placing the specimens in close contact with uranium and subjecting the assembly to reactor thermal neutrons. We may achieve this either by evaporating uranium onto the specimens or by placing them in contact with or close to a uranium foil. The latter method is preferable owing to the difficulties involved in protecting a vacuum system from contamination by radioactive material, and since a surface layer of uranium on the specimen would retain a considerable intensity of residual radioactivity following reactor exposure. Also, the specimens would have a film of foreign matter which might obscure or hinder interpretation of results. Thus a uranium foil which may be removed and disposed of after irradiation is the safest and most efficient source of fission fragments.

It is necessary to decide which isotope of uranium, U^{235} or U^{238}, will have dominance in the foil, and the choice depends first of all on the fission fragment intensity we desire, and secondly, therefore, as we shall shortly see, on the geometry of the experimental arrangement. 'Natural' uranium (containing mostly U^{238}) is some two orders of magnitude less efficient than 'enriched' uranium (containing mostly U^{235}) in undergoing thermal neutron-induced fission. Consider as an example an enriched foil whose thickness is 0·001 in., containing 93% of U^{235} and subjected to a thermal neutron flux of $1·2 \times 10^{12}$ neutrons/cm^2 sec (a typical experimental arrangement). The flux at the surface of the foil is in the region of $6·0 \times 10^9$ fragments/cm^2 sec, whereas if the foil

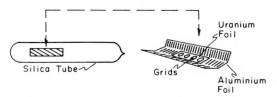

Figure 11. Experimental arrangement for irradiation with fission fragments in the reactor by direct contact of the specimen with uranium foil.

were of natural uranium the flux under identical conditions would be $\sim 4·5 \times 10^7$ fragments/cm^2 sec. A simple calculation shows that a 10 min exposure to this neutron flux produces an average of about three fission fragments per square of side 100 Å on the foil surface in the case of enriched foil and about three fragments per 1000 Å square with natural foil. Therefore, if we place an electron microscope grid in close contact with the foil, this flux will be incident on the specimens contained on the grid. For electron microscopy the total dose of fission fragments incident on the specimens during any feasible exposure time in the reactor is often excessively large when enriched uranium is used as a source, and for this reason natural foil is generally employed in the direct contact method of irradiation.

The experimental arrangement for irradiation by direct contact with a uranium foil of about 10 μ thickness is shown in Figure 11. The specimens are placed either directly on the foil or on electron microscope grids which are in turn laid on the foil. The whole is then wrapped in aluminium foil, sealed in a silica tube and enclosed in a metal cylinder which is lowered into the reactor core for the specified time (normally in the region of 10 min).

There is one major disadvantage of this arrangement, which is of particular significance if we are hoping to produce fission fragment tracks in a suitable specimen. The fission fragments leaving the uranium foil are to a first approximation randomly directed, and consequently very few fragments enter the specimen at low enough angles to form tracks of 1 μ or longer. Most of the damage takes the form of short tracks and spots which tend to hinder the electron microscope analysis of the few longer tracks which occur. Furthermore, the determination of the direction of travel of a fragment causing a particular long track involves a complicated procedure of image analysis (see Chapter 6). To overcome this difficulty, a technique resulting in collimation of the fission fragments was introduced by Bowden and Caspar (1963). Here

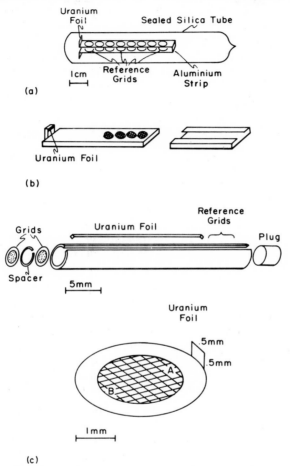

Figure 12. Techniques employed to irradiate specimens with a 'collimated' beam of fission fragments. (a) Employing adhesive, to both specimen and uranium. (b) Employing adhesive to uranium only. (c) Employing no adhesive.

the geometry of the irradiation is altered by separating the fission fragment source from the specimens by distances of the order of 1 or 2 cm, so that the solid angle presented by the source at the specimen is small. Several methods of achieving this are illustrated in Figures 12(a), (b), and (c). In (a), grids containing the specimens are affixed by a small part of their rims, using epoxy-resin, to a strip of aluminium ribbon bent in the shape shown. Uranium foil is similarly glued to the extremities of the strip, and the apparatus is then sealed in a silica tube under vacuum, since the range of fission fragments in air is only a few centimetres. With this geometry the fragments incident on a particular specimen are collimated to within 10° in the best case, the angle between the direction of fragment motion and the crystal surface being dependent on the distance between the crystal and uranium foil.

Although the collimation in this type of configuration is very good, it has the disadvantage that the specimens are subject to contamination by the small amounts of glue or epoxy-resin used to stick the grids to the aluminium. This difficulty is overcome in the modification shown in Figure 12(b), used by Morgan (1966). Specimen grids are placed in circular recesses cut into the aluminium strip, and a second, shaped fitting slides matchbox-fashion over the first, preventing the grids from falling out of their recesses. Uranium foil is glued to the end of the strip as before.

The third type of holder for collimated fission fragment irradiation, illustrated in Figure 12(c), requires no glue or adhesive, and was employed by Nordén (1965). It consists of an aluminium tube about 2 cm in length whose inner diameter is approximately the diameter of the specimen grids (~ 3 mm). A groove 0·5 mm wide is cut in the tube and fitted aluminium slugs stop-up the ends. The specimen grids and aluminium wire spacers are stacked inside. The groove holds the fission fragment source, in the form of a thin ribbon of uranium, and aluminium foil is wrapped around the assembly. Figure 12(c) also shows the geometry of the collimation in this type of holder, which allows us to compare different areas of each specimen under different conditions of flux and angle of bombardment. Hence, considering two points A and B on a grid, the maximum angle of incidence of the fragment is 40° at A and 10° at B, and the flux at A is 200 times the flux at B.

When we wish to study the damage caused by fission fragment bombardment in a specimen using high-resolution field-ion microscopy, a different experimental assembly is required for the irradiation. The specimens take the form of very fine (100 to 500 Å) tungsten tips mounted as they would be in the field-ion microscope. A specially designed holder enables these tips to be bombarded by fission fragments collimated to within 7°. The holder and the results of a field-ion microscope study of fission fragment damage are described in detail in Appendix 1.

In all irradiations where collimated fission fragments are incident on the specimen, the collimation is achieved at the cost of a reduced particle flux. Since only a small proportion of the fission fragments emanating from the uranium pass through the specimen, it is necessary either to subject the foil to prolonged exposure to the neutron flux, or to use enriched uranium. The latter course is the more desirable both for convenience and to reduce any masking effects of a lengthy neutron bombardment of the specimen. It is also advisable as a precautionary measure to include in each irradiation some specimens shielded from the flux of fission fragments emerging from the foil, in order to monitor the background effects of neutron bombardment, reactor fission fragments, and heating.

The Simulation of Fission Fragments by Beams of Heavy Ions

It has been clearly indicated in Chapters 1 and 2 that, while the fission fragment represents a conveniently obtained heavy particle of high energy and charge, we can say very little about a particular fragment which causes damage in a crystal. The only guides we possess are the statistical distributions of mass and energy and approximate determinations of charge variations. It is therefore desirable to obtain charged particles comparable to fission fragments, whose properties we more accurately know. This has been achieved by Datz, Noggle, and Moak (1965) through the use of bromine (^{79}Br) and iodine (^{127}I) ions, accelerated to energies of 20 to 80 MeV in a tandem accelerator.

The principle is to accelerate ions from a source to 7 MeV in a Van de Graff machine, at which point they pass through a stripper gas, emerging with a charge of $4e^+$, $5e^+$ or higher. The high charge characteristic of the fission fragment is obtained by keeping the gas pressure in the accelerator tube sufficiently high so that some ions will continuously lose electrons by collision, attaining charges up to $20e^+$. Since high energies and high ionic charges are simultaneous particle characteristics, the magnetic rigidity of each energy regime of the beam is constant, and a magnetic analyser in the beam path will precisely define the energies at which different charge states of ions pass at a particular setting of field. Thus, if the magnetic deflection constant is found by calibration (for example in a proton resonance experiment), the emerging heavy ion beam energies for the different charge states are accurately known. These energies may be altered simply by changing the field of the analyser.

The heavy ion beam obtained in this way is within the fission fragment spectrum of mass, energy, and charge, but may be used in more quantitative studies of heavy charged particle damage, since the characteristics of the ions which simulate fission fragments are more accurately determined than those of the fragments themselves.

In this brief summary of experimental irradiation techniques we have concentrated on the methods used to produce fission fragment damage in specimens of the material under study, which are generally applicable to the different classes of solid we shall consider shortly. Other facets of the experimentation, both pre- and post-irradiation, are localized to a particular substance or type of investigation, and will be described in appropriate chapters. Before considering the different types of fission fragment damage observed in solids, however, we discuss the use of high-resolution electron microscopy to study defects in crystals, showing how it has become an extremely valuable tool in revealing the damage caused by an individual fission fragment in its passage through a thin crystal.

References

BOWDEN, F. P., and CASPAR, P. E. (1963), *Phil. Mag.* **8**, 2091.
DATZ, S., NOGGLE, T. S., and MOAK, C. D. (1965), *Phys. Rev. Lett.* **15**, 254.
MORGAN, D. V. (1966), Ph.D. Thesis, Univ. of Cambridge.
NORDÉN, H. (1965) Ph.D. Thesis, Univ. of Cambridge.

CHAPTER FOUR

Direct Observation of the Bombarded Lattice

The observation of fringe patterns on thin crystals in the electron microscope has, since results of the first experiments were published by Menter (1956), been the object of considerable practical and theoretical attention. Fringe patterns can in general be due to several distinctly separate mechanisms. Crystal defects, for example, have been shown to give rise to periodic intensity undulations in the electron microscope image of a crystal. Whelan and Hirsch (1957a, b) have shown that stacking faults fall into this category, and Hirsch, Howie, and Whelan (1960, 1961) indicate the same mechanism to apply for inclined dislocation lines imaged by 'diffraction contrast'. The fringes which arise from a wedge-shaped crystal are essentially a thickness effect and extinction contours are due to long-range strains within the lattice. Perfect crystals produce two other types of pattern. When one crystal lies on top of another, double diffraction can lead to rotational or parallel type moiré patterns in the region of overlap. Alternatively, when only one crystal is imaged, it is possible to observe a system of fringes arising from direct inter-ference of the electron beam with the lattice itself, providing that certain stringent criteria are satisfied. The investigations reported in this chapter concern the latter type of fringe system unless otherwise stated.

The number of crystal types whose lattice fringes have been observed is now quite extensive. In the first instance Menter (1956) imaged the $(20\bar{1})$ planes in platinum and copper phthalocyanine $(C_{32}H_{16}N_8Cu)$ and obtained micrographs of fringe systems with a separation of about 10 Å. Labaw and Wyckoff (1957, 1958) and Chadderton (1960) made similar observations of other organic crystals, while Basset and Menter (1957), using the electron microscope at the highest resolution, showed that it was possible to photograph the (020) planes of molybdenum trioxide which have a separation of 6·93 Å. Observations of the molecular array have also been made with some protein crystals, and, in particular, Bradley (1959 and private communication) has made a careful and systematic examination of Finback Whale (type F) myoglobin crystals. Perhaps the most exciting observations, however, are those due to Saulnier and Mirand (1960), who have obtained electron micrographs of (220) planes in thin films of the face-centred-cubic superlattice βAl_3Mg_2, and Komoda (1966), who has imaged the lattices of aluminium, copper, and gold. In all these cases the fringe patterns possess three intrinsic properties. They have a contrast profile; they have a separation; and they have a direction which is able to be related to the morphological features of the crystal under examination. Papers by Labaw (1960), Chadderton (1960, 1961), Cowley and Moodie (1959c, 1960a, b), and Hashimoto, Mannami, and Naiki (1961) indicate that in the majority of cases the fringe separation and direction is directly associated with a particular set of crystal lattice planes and that in all cases there is no simple relation between the contrast profile and the potential distribution of the lattice itself. Notwithstanding the latter limitation, it was considered likely that a study of crystal lattices which had been irradiated with fission particles might well throw some light on the nature of the damage produced and possibly on the mode of formation of extended defects, such as dislocations, from point defects such as interstitials and vacancies. This chapter presents the details and results of experiments which have been conducted with this end in view.

Platinum phthalocyanine is, as Menter (1956) first showed, a substance which most nearly satisfies the special conditions which are required for experiments of this kind. In the first instance, all of the phthalocyanines show a surprising thermal stability, and Bowden and Chadderton (1961) have indicated that, even under the influence of quite high beam currents in the electron microscope, decomposition is a comparatively slow process. Secondly, the crystal structure of the phthalocyanines has been worked out by Robertson (1934, 1935) and by Robertson and Woodward (1937, 1940). Platinum phthalocyanine, for example, is monoclinic with a space group of $P2_{1/a}$ and a large β angle ($a_0 = 23 \cdot 9$ Å, $b_0 = 3 \cdot 87$ Å, $c_0 = 16 \cdot 9$ Å, $\beta = 129 \cdot 6°$). Moreover, the $(001)\overset{\wedge}{(}201)$ angle is 88°, so that the flat molecules of platinum phthalocyanine ($C_{32}H_{16}H_8Pt$) lie in almost vertical $(20\bar{1})$ planes when crystallization is such that the basal plane is on (001). Growth of electron-transparent platinum phthalocyanine crystals with this orientation is a comparatively easy process. Copper phthalocyanine, on the other hand, is often obtained in several polymorphic forms, and crystals with different habits grow under similar conditions. This can make the interpretation of fringe patterns in electron micrographs of crystals somewhat ambiguous. Hamm and van Normann (1948) and Suito and Uyeda (1957) have reported on habit modifications in copper phthalocyanine.

Fringe patterns on fission fragment irradiated crystals of platinum phthalocyanine have been shown to possess certain defects, and an interpretation is suggested in the following. A limited theoretical discussion is included and also a consideration of the changes in certain optical properties of single crystals of both platinum and copper phthalocyanine on irradiation with fission fragments.

Experimental

Platinum phthalocyanine was prepared from platinous chloride and phthalonitrile in the manner described by Barret, Dent, and Linstead (1936), and the material was purified by repeated vacuum sublimation. Crystal specimens for the electron microscope were obtained from the vapour phase by direct condensation on to the specimen support film. Figure 13 shows how crystals of

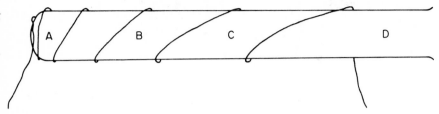

Figure 13. Apparatus used to prepare crystal specimens of platinum and copper phthalocyanine. The polycrystalline sample is placed at A. On heating *in vacuo*, or in a 10 cm pressure of argon, the temperature gradient is such that crystals of decreasing size are formed between B and C. Those crystals at B are suitable for optical studies and those at C are ideal for electron optical work. Inorganic impurities remain at A and organic impurities collect at D.

different size and thickness were prepared and Figure 14 depicts the normal habit of the crystal. The specimen support films have been either a perforated carbon layer or else formvar nets prepared by using an improved method due to Sjöstrand (1956). Crystals obtained by the above process are sometimes

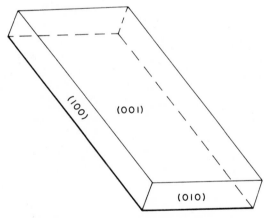

Figure 14. Predominant habit of the crystals formed between B and C in the apparatus shown in Figure 13.

deposited over the holes; the contrast is enhanced and there is no underlying film to distort the image of the lattice. Very thin crystals of platinum phthalocyanine are shown in Figure 15.

All the specimens were placed, on the support films, against natural uranium foil of about 10 μ thickness and irradiated in the pile BEPO at Harwell for times from a few minutes up to a few hours at a thermal neutron flux of

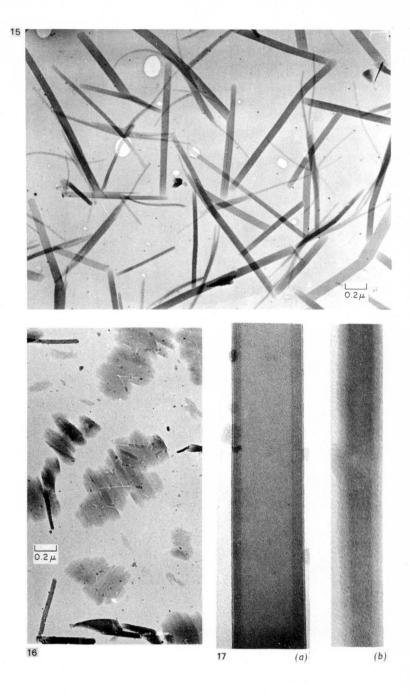

15

16

17   (a)   (b)

0.2 μ

0.2 μ

approximately $1 \cdot 2 \times 10^{12}$ neutrons/cm^2 sec. Identical specimens, except that they lacked the uranium foil, were irradiated with thermal neutrons only, for the same time and at the same flux. This served as a control experiment. Because of the small quantities used, the only troublesome remaining activity in the specimens was due to the ^{64}Cu of the supporting grids. This decayed by β-emission with a half-life of $12 \cdot 8$ h, so that after several days it proved possible to examine the specimens in the electron microscope. A Siemens Elmiskop 1 instrument operating at an accelerating voltage of 80 kV ($\lambda = 0 \cdot 0417$ Å) was used throughout. A 200 μ condenser aperture and 50 μ objective aperture were most often employed and the electron-optical assembly was always rigorously and carefully alined. Particular attention was always paid to the proper adjustment of objective lens astigmatism. The absence of both astigmatism streaks and specimen drift generally permits a healthier interpretation of high-resolution electron micrographs. Other precautions for using the electron microscope at the optimum resolution have been listed by Haine and Mulvey (1954). The images were recorded on Ilford Special Contrasty Lantern plates at a magnification of 80,000. Long exposures (45 sec) were necessary since the electron beam intensity was low in order to reduce both the effect of the electrons on the crystal and the effect of contaminating material in the vacuum system.

A Unicam SP 500 absorption spectrometer was used to determine the absorption spectra of single copper and platinum phthalocyanine crystals before and after irradiation. For this purpose thin, flat, crystalline ribbons of the copper derivative were grown by entrainer-gas vacuum sublimation in a sealed silica tube in a furnace. At temperatures between 600 and 700°C and pressures of about 10 cm of argon or nitrogen, beautiful thin crystals of copper phthalocyanine in the (001) orientation were formed. Thicknesses ranged from about one-fourth of a micron up to several microns and the crystals were blue when observed in transmission. Specimens were irradiated with fission fragments in the manner already discussed and the absorption spectra obtained using a cryogenic Dewar.

Results

Electron Microscopy

When crystals irradiated with both fission fragments and thermal neutrons were examined at high magnification in the electron microscope, a network of tracks and spots of light contrast was revealed. This is shown in Figure 16. No

---

Figure 15. Crystals of platinum phthalocyanine as imaged at low magnification in the electron microscope. Some of the crystals lie over holes in the supporting film.

Figure 16. Electron micrograph of fission fragment irradiated crystals of platinum phthalocyanine. Tracks and spots are present in the crystals.

Figure 17. High resolution electron micrographs of crystals of platinum phthalocyanine showing the perfect structure of the (20$\bar{1}$) fringes before irradiation with fission fragments. (a) Bands of higher contrast at the crystal edge indicate the presence of growth steps. (b) A very thin crystal which is only some nineteen molecules in width.

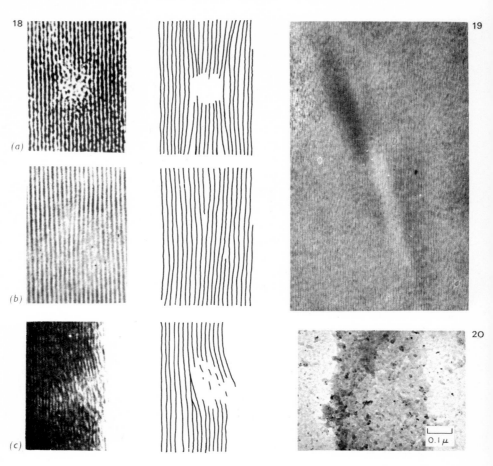

Figure 18. (a) Electron micrograph of a platinum phthalocyanine crystal showing the molecular disarray when a fission fragment travels perpendicular to the projection of the (20$\bar{1}$) planes and within the main body of the crystal. The drawing illustrates the situation more clearly. (b) Electron micrograph of a platinum phthalocyanine crystal showing the damage sustained when a fission fragment travels perpendicular to the projection of the (20$\bar{1}$) and within the main body of the crystal. A dislocation dipole results. (c) Electron micrograph of a platinum phthalocyanine crystal showing the damage sustained when a fission fragment strikes a crystal edge.

Figure 19. Electron micrograph of a fission fragment track in a crystal of platinum phthalocyanine.

Figure 20. Electron micrograph of a crystal of platinum phthalocyanine which has sustained a dose of fission fragments in excess of 10^{13} particles/cm^2. The former crystallinity is destroyed and only amorphous carbon remains.

such network was observed in crystals irradiated with thermal neutrons only, and at high magnification the fringe pattern deriving from the $(20\bar{1})$ planes was perfect (Figure 17). In the fission fragment irradiated crystals, however, the fringe pattern was no longer perfect. At the spots, several kinds of fringe distortion were observed and examples are reproduced in Figure 18. Figure 18(a) shows how in some cases the $(20\bar{1})$ fringes curve inwards towards a diffuse region some seven fringes in diameter. The situation is similar in Figure 18(b), but the diffuse region no longer exists and it is evident that two abruptly terminating fringes are present. In Figure 18(c) the large-scale fringe distortion at a crystal edge is shown. No such distortions were observed at or near the tracks of light contrast. In a very few cases the tracks seemed merely to cut the crystal into two parts separated by about 100 Å, but the great majority, as in Figure 19, were merely linear contrast effects apparently superimposed upon a perfect $(20\bar{1})$ fringe pattern. Figure 19 also shows another phenomenon which was frequently observed; the sudden reversal and side-stepping of contrast at discrete points along the line.

After a very heavy dose of fission fragments of about 18 h duration no spots or tracks were observed. The crystals took on an ill-defined granular structure (Figure 20) and the electron transmission diffraction pattern was that of amorphous carbon only.

Optical Properties

It proved difficult to prepare large quantities of thin platinum phthalocyanine crystals so the results reported in this section refer, in the main, to copper phthalocyanine. The absorption spectrum of the one good platinum phthalocyanine crystal obtained is shown in Figure 21(a). Unpolarized light was used and the dichroism of the phthalocyanines was not investigated. The close similarity with the absorption spectrum of copper phthalocyanine shown in Figure 21(b) is immediately obvious.

Changes in the absorption spectra of copper and platinum phthalocyanine on irradiation with fission fragments and thermal neutrons are given in Figure 22. No such changes were observed in copper phthalocyanine irradiated with thermal neutrons only.

Interpretation of Results

There can be little doubt that the tracks and spots observed in the electron microscope on thin crystals of platinum phthalocyanine are due to the passage through the lattice of individual fission fragments. Similar features have been observed in other materials treated in a like manner. In particular, Silk and Barnes (1959) and Bonfiglioli, Ferro, and Mojoni (1960) have detected fission fragment tracks in mica, while Izui and Fujita (1961) have drawn attention to linear arrays of strain centres in fission fragment bombarded graphite. Bierlein and Mastel (1961) have made a detailed study of several polycrystalline films of different thicknesses but have extended their experiments to thin aluminium single crystalline foils. Noggle and Stiegler (1960) have been concerned with thin films of uranium dioxide irradiated with thermal neutrons, and Bowden and Montagu-Pollock (1961) have described the damage produced by a fission

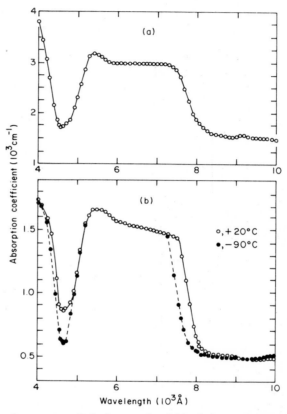

Figure 21. Absorption spectra of single crystals of (a) platinum phthalocyanine and (b) copper phthalocyanine. Unpolarized light was used.

fragment in an explosive crystal, silver cyanamide. All the authors listed conclude that tracks and spots are directly due to individual fission fragments. No comparable phenomena are observed in specimens irradiated with thermal neutrons alone.

The number of fission fragments entering the specimen in unit time is dependent on the thermal neutron flux in the reactor, on the geometry of the experimental arrangement used, and on the thickness and constitution of the uranium foil. For a thermal neutron flux of $1·2 \times 10^{12}$ neutrons/cm²/sec in BEPO and a natural uranium foil thickness of 10 μ, the fission fragment flux, with the geometrical arrangement used, was estimated to be about $5·7 \times 10^7$ fragments/cm²/sec. This figure is in sufficiently good agreement with the

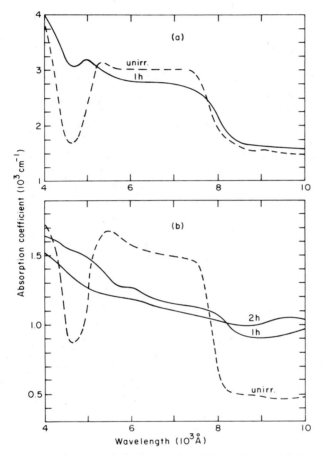

Figure 22. Absorption spectra of single crystals of (a) platinum phthalocyanine and (b) copper phthalocyanine. Flux of fission fragment irradiation, approximately $5·7 \times 10^7$ fragments/cm² sec.

observed density of tracks and spots for specimens of platinum phthalocyanine irradiated for different times, and the number of tracks and spots is directly proportional to the exposure time in BEPO. It can be assumed that each fission particle produces its own localized contrast effect in crystals of platinum phthalocyanine.

The one-to-one correspondence between the number of particles impinging on the crystal and the number of disturbances produced implies that both tracks and spots are due to individual fragments. It therefore seems probable that the inclination of the path of the recoil atoms born in the fission event with the plane of the crystal ribbon lies between two extremes. The fission fragment may pass vertically through the crystal producing a spot or it may travel at a low angle to the (001) plane and hence produce a track. Tracks with lengths of up to 5 μ have been observed. The fact that in general both lines and spots possessed a light contrast suggests that, in platinum phthalocyanine, such areas are more transparent to electrons than the remainder of the crystal. Less matter may be present in these regions. The disturbances sustained by the (20$\bar{1}$) lattice fringes when a spot is examined at high resolution are most interesting. The points of special interest may be listed:

(1) The fragment produces a diffuse area of light contrast which extends in diameter to about eight lattice fringes or 100 Å (one lattice fringe = 11·94 Å). At the edge of this area there is a marked inward curvature of the (20$\bar{1}$) lattice fringes (see Figure 18(a)).

(2) The fragment produces an area of light contrast which extends in diameter up to about 100 Å but which is no longer diffuse. Two half-fringes, separated by usually four or five perfectly continuous fringes, are observed to end abruptly in the area of damage. There is a marked curvature of the (20$\bar{1}$) lattice fringes adjacent to the area of damage (see Figure 18(b)).

(3) The fragment impinges near a crystal edge and produces a distortion of the lattice fringes which can extend over as much as 250 Å (see Figure 18(c)).

The classical theoretical papers of Cowley and Moodie (1957a, b, c, 1959a, b, c, 1960a, b) have shown that it is necessary to exercise extreme care in interpreting lattice fringe patterns observed in the electron microscope. Even when the specimen lattice is perfect, the resulting fringe pattern, arising from interference of the electron waves with the three-dimensional repetitive array of molecules, represents the lattice in only an indirect way. For the purpose of this chapter it is sufficient to consider only very thin crystals, since that is the type of crystal which has been used in the above experiments. In such circumstances the platinum phthalocyanine crystal can be considered to behave as a periodic *phase* object. If the electron microscope were focused accurately on the exit face of the crystal (a condition it is impossible to obtain in practice) then the contrast would arise only from inelastically scattered electrons which suffer considerable energy losses. For *thin* crystals, however, the inelastic scattering is much less than the elastic scattering, so the contrast would be small and negligible. Zero contrast would also be obtained on a series of planes lying parallel to, and above and below the exit face of the crystal. These are the so-called Fourier image planes of Cowley and Moodie (1957a, b, c, d, 1960a, b). At all the intermediate 'out-of-focus' positions, a fringe pattern would be obtained. The introduction of a 50 μ objective lens aperture into the electron optical assembly introduces contrast patterns at the Fourier image planes and modifies the intermediate patterns. Discussion of these effects is deferred.

Interpretation of Fringe Patterns

It is pertinent to ask to what extent the fringe disturbances seen in fission fragment irradiated crystals of platinum phthalocyanine can be considered to represent the distortion of the $(20\bar{1})$ molecular planes. The effect of a potential field of $\phi(xyz)$, for a very thin crystal slice acting as a phase grating, is given by multiplying the wave function $\psi_0(xy)$, representing the incident wave, by the transmission function:

$$q(xy) = \exp\left[i\sigma\phi(xy)\right] \tag{4.1}$$

where

$$\phi(xy) = \int_{-\infty}^{+\infty} \phi(xyz)\,dz \tag{4.2}$$

z is the direction of the electron beam,

$$\sigma = 2\pi m\lambda/h^2 \tag{4.3}$$

and λ and m are the electron wavelength and mass, respectively.

The lattice itself may be represented by a general two-dimensional Fourier series:

$$\phi(xy) = \sum_n \sum_m C_{nm} \exp\left[2\pi i\left(\frac{nx}{a} + \frac{my}{b}\right)\right] \tag{4.4}$$

Cowley and Moodie (1960b) have shown that, in these circumstances, the wave function on an arbitrary plane of observation is given by:

$$\psi(xy) = \sum_n \sum_m C'_{nm} \exp(i\theta)\exp\left[2\pi i\mu(n^2 + m^2)\right] \tag{4.5}$$

where

$$\theta = 2\pi\left[\frac{nx}{a} + \frac{my}{b}\right] = \frac{2\pi}{a}(nx + my) \quad \text{(if } a = b) \tag{4.6}$$

$$\mu = \frac{\lambda_0}{2a^2}\frac{RR_q}{(R + R_q)} \tag{4.7}$$

λ_0 is the electron wavelength *in vacuo*, and R and R_q are the distances of the object from the observation plane and the source, respectively.

C'_{nm} are the Fourier coefficients of $q(xy)$ if this is represented as a two-dimensional series, and values of μ characterize the planes of observation. If μ is half-integral or integral, the Fourier image planes are defined; the intensity is a constant and no contrast due to the crystal lattice structure is present. Fringe patterns do arise, however, on planes a small distance from the Fourier image planes. Thus, if the plane defined by the relation:

$$\mu = (n + \delta) \quad (n \text{ an integer}) \tag{4.8}$$

is considered, and if the inequality:

$$\delta(n^2 + m^2) \ll 1 \tag{4.9}$$

is valid for all n and m, then the intensity of the contrast on that plane is given by:

$$I = 1 + \tfrac{1}{2}(l\sigma\lambda/\pi)\,\nabla^2\,\phi(xy) \tag{4.10}$$

Similarly, on planes midway between Fourier image planes, defined by μ being quarter-integral, fringe patterns are formed:

$$I_{\mu=1/4} = 1 + \sin 2\sigma\phi_{oe}\sin 2\sigma\phi_{eo} - \cos 2\sigma\phi_{oo}(\sin 2\sigma\phi_{eo} + \sin 2\sigma\phi_{oe}) \tag{4.11}$$

$$I_{\mu=-1/4} = 1 + \sin 2\sigma\phi_{oe}\sin 2\sigma\phi_{eo} + \cos 2\sigma\phi_{oo}(\sin 2\sigma\phi_{eo} + \sin 2\sigma\phi_{oe}) \tag{4.12}$$

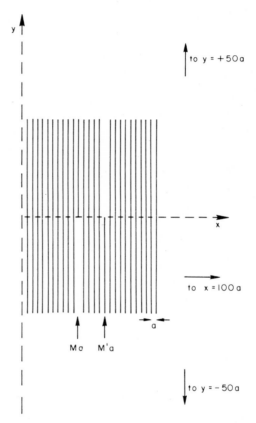

Figure 23. Diagram to show the variations in the xy-plane if the function (xy) represents a crystal with a dislocation dipole (equation 4.14).

In these equations, functions ϕ_{oe} are of the form:

$$\phi_{oe} = \sum_{n\ \mathrm{odd}}\ \sum_{m\ \mathrm{even}} C_{nm}\exp\{i\theta\} \tag{4.13}$$

Consider, now, the effect of assuming certain distortions to be present in the molecular $(20\bar{1})$ planes and determine the effect on the fringe patterns – where these are observable. In the first instance let it be assumed, as might be suspected from the second type of fringe disturbance listed above, that the effect of the fission fragment on the lattice had been to introduce two extended imperfections characterized by two oppositely directly half-planes. Ideally, this may be represented mathematically by a function of the type:

$$\phi(xy) = A\sin^2(\pi x/a)\{1 - \theta(-y)[\theta(x - Ma) - \theta(x - Ma - a)]$$
$$- \theta(y)[\theta(x - M'a) - \theta(x - M'a - a)]\} \qquad (4.14)$$

This is shown in diagrammatic form in Figure 23, for a crystal with a cubic unit-cell, side length a, and with a width and length of one hundred unit-cells. If the same function is expressed in the form of a two-dimensional Fourier series:

$$\phi(xy) = \sum_{-\infty}^{\infty} \sum_{-\infty}^{\infty} C_{nm} \exp\left(\frac{2n\pi\,ix}{100a}\right)\exp\left(\frac{2m\pi\,iy}{100a}\right) \qquad (4.15)$$

then the Fourier coefficients are given by:

$$C_{nm} = \frac{1}{10000a^2}\int_0^{100a} dy \int_0^{100a} \phi(xy)\exp\left(-\frac{2n\pi\,ix}{100a}\right)\exp\left(-\frac{2m\pi iy}{100a}\right)dx \qquad (4.16)$$

Hence, for the function $\phi(xy)$ given in equation 4.14, it may be shown that the corresponding Fourier coefficients have the form:

$$C_{nm} = \frac{1}{10000a^2}\{[A\chi(x)]_0^{+100a}[\Phi(y)]_{-50a}^{+50a} - [A\chi(x)]_{Ma}^{(M+1)a}[\Phi(y)]_{-50a}^0$$

$$-[A\chi(x)]_{Ma'}^{(M'+1)a}\ [\Phi(y)]_0^{+50a}\} \qquad (4.17)$$

The lattice disturbance therefore asserts its presence in the coefficients of the Fourier series. The effect of such disturbances on the fringe patterns obtained on planes at different out-of-focus positions may now be considered. For a plane only a small distance l away from a Fourier image plane (at which contrast is zero) equation 4.10 is applicable.

Now, for electrons:

$$\nabla^2\phi = 4\pi\rho(xy) \qquad (4.18)$$

where $\rho(xy)$ is a projection of the charge distribution. Hence:

$$I = 1 + 2l\sigma\lambda\rho(xy) \qquad (4.19)$$

so it may be immediately deduced that the lattice disturbance is directly imaged at such planes of observation. Consider further those planes, lying midway between Fourier image planes, which are characterized by quarter-values of μ. In particular, for $\mu = \frac{1}{4}$, equation 4.11 applies. If consideration is restricted to small crystal thicknesses and small phase changes, equation 4.11 may be written:

$$I_{\mu=1/4} = 1 - 2\sigma(\phi_{oe} + \phi_{eo}) \qquad (4.20)$$

and, since the functions ϕ_{oe} are expressible in terms of the Fourier coefficients C_{nm} of the function $\phi(xy)$ in equation 4.14, the lattice disturbance is also again observed in the fringe pattern. It is to be concluded that, under the ideal conditions stipulated, a disturbance of the lattice resulting from irradiation with fission fragments, including a curvature of the molecular planes (as listed in (1) and (2) above), is likely to be present in the interference patterns observed on planes characterized by:

$$\mu = n + \delta \qquad (4.21)$$

and

$$\mu = \tfrac{1}{4}(2n + 1) \qquad (4.22)$$

It is important to point out that the conditions under which the above equations apply are ideal. No account has been taken of aberrations in the electron-optical lenses and the effect of the introduction of an objective lens aperture has so far been neglected. Cowley and Moodie (private communication) have indicated how the objective aperture may be accounted for and their argument lends itself admirably to the present discussion. It is assumed that the aperture, when this is introduced into the microscope assembly, discovers only a limited number of the diffraction spots which are formed in the back-focal plane of the objective lens, and these are grouped around the primary spot. In general, a step-like function will introduce the aperture effect into the mathematics but, in particular, the generating function for Bessel coefficients may be invoked. Thus, equation 4.1 may equally well be written:

$$q(xy) = \sum_{-\infty}^{\infty} \{i\phi \pm \sqrt{(1 - \phi^2)}\}^n \mathrm{J}_n(\sigma) \qquad (4.23)$$

and, if only the primary spot and its two immediate neighbours on the central layer line of the diffraction pattern are revealed by the aperture, the wave function at the plane of observation becomes:

$$\psi_\mu(xy) = \mathrm{J}_0(\sigma) + [\{i\phi \pm \sqrt{(1 - \phi^2)}\} \mathrm{J}_1(\sigma)$$
$$- \{i\phi \pm \sqrt{(1 - \phi^2)}\}^{-1} \mathrm{J}_1(\sigma)] \exp(2\pi \, i\mu) \qquad (4.24)$$

since

$$n = 0, \pm 1 \qquad (4.25)$$

The intensity variations on this plane of observation may then be derived as follows:

$$I_\mu(xy) = \mathrm{J}_0^2(\sigma) - 4\mathrm{J}_0(\sigma)\,\mathrm{J}_1(\sigma)\,\phi \sin(2\pi \, \mu) + 4\mathrm{J}_1^2(\sigma)\,\phi^2 \quad \text{(for the positive root)}$$
$$(4.26)$$

The contrast patterns on Fourier image planes will have a periodicity given by squaring the function ϕ, but for σ small and small crystal thicknesses the term in ϕ dominates at intermediate positions. Disturbances in the crystal lattice are therefore present in the fringe patterns observed at these intermediate positions. In this treatment a general case has been considered but similar equations may be obtained for particular crystals whose lattice is able to be 'imaged' in the electron microscope. Lens aberrations, which may be important when the microscope is operated at high magnification and resolution, have not been discussed.

The Damaged Lattice

Since the fringe patterns observed in the experiments reported here have mostly been obtained at out-of-focus positions, and since the periodicity has always been that of the (20$\bar{1}$) planes themselves, it may be concluded that the patterns closely represent the lattice in both spacing and direction. The three points of special interest for irradiated thin crystals of platinum phthalocyanine should now be considered again.

(1) The fragment produces an area of damage which extends in diameter up to about eight lattice planes. At the edges of this area there may be a marked inward curvature of the (20$\bar{1}$) molecular planes.

(2) The fragment produces an area of damage which *may* be considered to be a pair of oppositely directed half-planes, separated by four or five perfect, but locally curved, continuous planes. Bollman (1961) has called such an arrangement a dislocation dipole. Two edge dislocations which are of opposite sign and which are situated on neighbouring parallel planes will not annihilate completely, though they can combine to form a row of vacancies. The state of affairs noted in (1) would then be observed, and the long-range stress field would be removed. Alternatively, depending on the separation of the two dislocation components in directions perpendicular and parallel to the lattice planes, it is possible for a repulsion to occur and even a partial self-destruction of the two components is then forbidden.

It has been pointed out by Basset, Menter, and Pashley (1958) that an extra half-fringe does not necessarily imply an extra half-plane of molecules lying perpendicular to the crystal surface. A dislocation may lie at a more general angle to the basal plane and still terminate on the upper and lower surfaces of the crystal. In such circumstances a half-fringe, or several half-fringes, or an ill-defined region of fringe discontinuity may result in the pattern observed. It is therefore not possible to indicate the Burger's vectors of the components of a dislocation dipole of the type described above.

(3) The fragment produces an area of damage at a crystal edge which is larger than that observed in regions remote from the edge of the crystal. The fringe pattern disturbance is exceedingly complex and an interpretation in terms of dislocations in the (20$\bar{1}$) or other molecular planes is not possible.

Emphasis should at this point be again laid on the necessity for exercising extreme caution in linking fringe discontinuities with dislocations. That this is of the utmost importance will be evident from the fact that both contamination and irregularities in the supporting film, if the latter is present, can lead to fringe disturbances which may include half-fringes. In the present work, however, there is a distinct correlation between light contrast spots, which are undoubtedly due to fission fragments, and fringe disturbances. Moreover, the necessity of having an underlying specimen support film has been avoided.

The observations made on fission fragment tracks in thin platinum phthalocyanine crystals suggest a disturbance of the crystalline material in a cylinder some 100 Å in diameter. In crystals less than 100 Å thick, therefore, the effect of a fragment travelling at an extremely low angle to the basal plane of the crystal would be to separate the latter into two parts. The occurrence of extended tracks in very thin crystals is quite rare – there are far more spots than tracks – but when these do occur the (20$\bar{1}$) fringes are apparently perfect in the two parts. This may mean that the (20$\bar{1}$) molecular planes are perfect in the two parts in regions adjacent to the track.

In thicker crystals (~ 400 Å) the tracks are merely linear regions of light contrast which are crossed by perfect (20$\bar{1}$) fringes. This suggests that the perfect crystal lattice which lies above and below the cylindrical track is clearly imaged, and that the projection of the disturbed cylinder is not sufficient to distort the image. The track contrast effect which has been observed in thick crystals (Figure 19) is explicable only in terms of the dynamic theory of electron diffraction (see Chapter 6). Hirsch *et al.* (1960, 1961) have given explanations of similar phenomena observed at inclined dislocations in metal foils. Hashimoto *et al.* (1961) have published a detailed analysis of dynamic effects in electron microscope images of platinum phthalocyanine crystals.

When there are so many spots and tracks that multiple over-lapping occurs, the former crystallinity is completely destroyed. Amorphous carbon remains and there is no trace of the platinum.

Optical Properties

The semiconductor properties of some of the phthalocyanine derivatives have stimulated interest in the photoelectric and optical properties observed in the same compounds. Fielding and Gutman (1957) have reported on the relation between the absorption spectra of thin sublimed films and the temperature-dependence of resistivity, while the Russian authors, Vartanyan and Karpovich (1956, 1958a, b), have been concerned with the spectral dependence of the photoconductivity and optical activation energy. Terenin, Putzeiko, and Akimov (1957) point out the importance of such studies because of the similarity of the phthalocyanine and chlorophyll molecules, and, more recently, Tollin, Kearns, and Calvin (1960, 1961) have published a detailed treatment of the kinetics of conductivity in metal-free phthalocyanine. Without exception these treatments all deal with sublimed layers. An investigation of the optical properties of single crystals is reported by Lyons, Walsh, and White (1960) who give the reflection absorption spectra of phthalocyanine.

A most reasonable explanation for the transmission absorption spectra of thin single crystals of phthalocyanine and its derivatives involves a consideration of the π-electrons in the conjugated double-bond systems (Chadderton, 1963). If a crystal is illuminated, or if energy is supplied in some other manner, electrons are lifted to an excited level in the same molecule and the type of semiconductivity and photoconductivity observed is dependent on the mode of communication of the energy to the surrounding crystal environment. There is evidence that, in crystal composed of small molecules, the electron and resulting hole travel together in the form of an exciton and the theory of molecular crystals presented by Davydov (1948) is then involved. For the larger molecules, such as phthalocyanine, the electron and hole may diffuse independently through the crystal and the wavelength dependence of the photoconductivity and optical absorption will then be closely similar. It has in fact been established (Chadderton, 1963) that for copper phthalocyanine the photoconductivity response curve closely follows the optical absorption spectrum for a single crystal. This also suggests that, for molecular crystals like copper phthalocyanine, the energy required to jump the π-electron across an intermolecular barrier is negligible compared with the energy required to lift the electron to the excited state. The quasi-continuous bands shown in Figure 21 for platinum and copper phthalocyanine can be correlated with sharper peaks observed in transmission absorption spectra of vapour, solution, and evaporated films. If the spectral shifts arising from temperature-dependence and the broadening due to crystal field interactions are accounted for, there appears to be a one-to-one correspondence between peaks observed in the absorption spectra for molecules in different degrees of association. In particular it would appear that a doubly degenerate $B_{1u} - E_g$ transition which has been predicted by Basu (1954) in a molecular orbital calculation on phthalocyanine is present in the sublimed film spectra for this compound. Moreover, this peak, at a wave number of 14720 cm^{-1} (1·82 eV), has a close parallel in the platinum and copper derivatives. It would appear that the simple band model offers a reasonably good description of the behaviour of crystalline phthalocyanines, though the low values of absorption coefficient in Figures 21 and 22 give cause for some concern. It is possible that indirect transitions may be involved.

An explanation of the changes induced in the absorption spectra of copper and platinum phthalocyanine on irradiation with fission fragments may now be attempted. It is evident from Figure 22 that, with increasing exposure times, the absorption bands broaden and become less uniquely defined. An increased overlap of the bands is evident in the region between 4000 and 6000 Å, and the absorption edges are not so sharp. A smearing out of the energy bands in a manner such as this could result from perturbations in the crystal field. These, in their turn, could be due to crystal defects, resulting from fission particle bombardment. Interstitial atoms, vacant lattice sites and *extended* defects such as dislocations may be the reason for such perturbations.

There is also evidence for crystal field effects in infrared absorption spectra of irradiated crystalline suspensions. Irradiation of copper phthalocyanine with fission fragments leads to a doubling of the bands at wave numbers of 720 and 769 cm^{-1} in the original spectrum. There can be little doubt that the structure, or at least the crystal environment, of a proportion of the benzene ring has been affected, as these two bands are probably both caused by 'out-of-plane' vibrations of the C—H bonds of the benzene ring. The stronger band at 720 cm^{-1} is certainly due to such vibrations.

The infrared spectra of pure phthalocyanine and its derivatives have been listed by Terenin and Sidorov (1956).

Conclusion

In platinum phthalocyanine it is probably reasonable to conclude that the bulk of the damage is due to the electronic excitation and ionization. Seitz and Koehler (1956) suggest that the peripheral atomic Coulomb collisions may be important for fission fragments and they point out that the regions along the track where the thermal spikes are significant may then merge. In either event the problem may be simplified by assuming that the path of the fission fragment is representable by a cylindrical thermal spike, in which the majority of the energy is delivered by the electrons (Chapter 2). If one applies the normal classical laws of lattice heat conduction and if it is assumed that a sudden liberation of heat energy in a localized area takes place, it is then possible to calculate the temperature distribution with time in a thermal spike. Calculations performed on this basis for crystals of platinum phthalocyanine indicate that temperatures of over $1000°K$ endure for periods of about 10^{-11} sec. Now platinum phthalocyanine sublimes at about $500°C$, so a degree of thermal agitation will result along the track of the fission fragment, though this disturbance may not have the configuration of a true vapour, owing to the confining effect of the surrounding crystal. Robinson and Klein (1957) and Lawton (1957) have reported on the remarkable stability of copper phthalocyanine to irradiation and to heat respectively and the same conclusions must be drawn from the present work. It is possible that some molecules will either be lost or decomposed and especially in the thinner crystals. A relaxation of the molecular planes could then lead to the situation imaged in Figure 18(a) while that shown in Figure 18(b), the dislocation dipole, may be due to a recrystallization which is not in accord with the surrounding lattice. There is certainly evidence for the production of dislocations, and the remarkable disturbances which are shown in Figure 18(c) are probably due to thermally induced plastic flow at the surface of the crystal where volume restraints will be less.

The observations which have been reported in this chapter demonstrate conclusively that dislocations can be formed when crystalline matter is bombarded with uranium fission fragments, and emphasis has been placed on the use of a direct observation of the lattice fringes in the electron microscope in order to reach this conclusion. Along the heated cylindrical track of the fragment the crystalline matter is disturbed, decomposed, or removed. The subsequent arrangement is not necessarily perfect and strain centres or dislocations remain.

References

BARRET, P. A., DENT, C. E., and LINSTEAD, R. P. (1936), *J. Chem. Soc.*, p. 1719.

BASSET, G. A., and MENTER, J. W. (1957), *Phil. Mag.* **2**, 1482.

BASSET, G. A., MENTER, J. W., and PASHLEY, D. W. (1958), *Proc. Roy. Soc. A.* **246**, 345.

BASU, S. (1954), *Indian J. Phys.* **28**, 511.

BIERLEIN, T. K., and MASTEL, B. (1961), USAE Comm. Contract No. AT(45-1)-1350.

BOLLMAN, W. (1961), *J. Appl. Phys.* **32**, 869.

BONFIGLIOLI, G., FERRO, H., and MOJONI, A. (1960), US Air Force ARDC Contract No. AF.61(514)-1333 (1960).

BOWDEN, F. P., and CHADDERTON, L. T. (1961), *Nature*, Lond. **192**, 31.

BOWDEN, F. P., and MONTAGU-POLLOCK, H. M. (1961), *Nature*, Lond. **191**, 550.

BRADLEY, D. E. (1959), *Nature*, Lond. **183**, 941.

CHADDERTON, L. T. (1960), *Europ. Reg. Conf. Electr. Microsc.*, Delft, Nederlandse Verenging voor Electronenmicroscopie (Delft, 1961).

CHADDERTON, L. T. (1961), *Nature*, Lond. **189**, 564.

CHADDERTON, L. T. (1963), *J. Phys. Chem. Sol.* **24**, 751.

COWLEY, J. M., and MOODIE, A. F. (1957a, b, c), *Proc. Phys. Soc.* **70**, 486, 497, 505.

COWLEY, J. M., and MOODIE, A. F. (1957d), *Acta Cryst.* **10**, 609.

COWLEY, J. M., and MOODIE, A. F. (1959a, b, c), *Acta Cryst.* **12**, 353, 360, 367.

COWLEY, J. M., and MOODIE, A. F. (1960a), *Europ. Reg. Conf. Electr. Microsc.*, Delft, Nederlandse Verenging voor Electronenmicroscopie (Delft, 1961).

COWLEY, J. M., and MOODIE, A. F. (1960b), *Proc. Phys. Soc.* **76**, 378.

DAVYDOV, A. S. (1948), *J. Expt. Theor. Phys.* (USSR) **18**, 210.

FIELDING, P. E., and GUTMAN, F. (1957), *J. Chem. Phys.* **26**, 411.

HAINE, M. E., and MULVEY, T. (1954), *Proc. Int. Conf. Electr. Microsc.*, London, p. 698.

HAMM, F. H., and van NORMANN, E. (1948), *J. Appl. Phys.* **19**, 1097.

HASHIOMOTO, H., MANNAMI, M., and NAIKI, T. (1961), *Phil. Trans. A* **253**, 459.

HIRSCH, P. B., HOWIE, A., and WHELAN, M. J. (1960), *Phil. Trans. A* **252**, 499.

HIRSCH, P. B., HOWIE, A., and WHELAN, M. J. (1961), *Proc. Roy. Soc. A* **263**, 217.

IZUI, K., and FUJITA, F. E. (1961), *J. Phys. Soc. Japan* **16**, 1032.

KOMODA, T. (1966), *Japan J. Appl. Phys.* **5**, 419.

LABAW, L. W. (1960), *J. Ultrastruct. Res.* **4**, 92.

LABAW, L. W., and WYCKOFF, R. W. G. (1957), *Proc. Nat. Acad. Sci. Wash.* **43**, 1032.

LABAW, L. W., and WYCKOFF, R. W. G. (1958), *Proc. Int. Conf. Electr. Microsc.*, Berlin, **1**, 344, Springer Verlag (Berlin, 1960).

LAWTON, E. A. (1957), *J. Phys. Chem.* **62**, 384.

LYONS, L. E., WALSH, J., and WHITE, J. W. (1960), *J. Chem. Soc.*, p. 167.

MENTER, J. W. (1956), *Proc. Roy. Soc. A* **236**, 119.

NOGGLE, T. S., and STIEGLER, J. O. (1960), *J. Appl. Phys.* **31**, 2199.

ROBERTSON, J. M. (1934), *J. Chem. Soc.*, p. 615.

ROBERTSON, J. M. (1935), *J. Chem. Soc.*, p. 1195.

ROBERTSON, J. M., and WOODWARD, I. (1937), *J. Chem. Soc.*, p. 219.

ROBERTSON, J. M., and WOODWARD, I. (1940), *J. Chem. Soc.*, p. 36.

ROBINSON, M. T., and KLEIN, G. E. (1957), *J. Phys. Chem.* **61**, 1004.

SAULNIER, A., and MIRAND, P. (1960), *Mem. Sci. Rev. Met.* **57**, 91.

SEITZ, F., and KOEHLER, J. S. (1956), *Solid State Physics* **2**, Academic Press (New York).

SILK, E. C. H., and BARNES, R. S. (1959), *Phil. Mag.* **4**, 970.

SJÖSTRAND, F. S. (1956), *Europ. Reg. Conf. Electr. Microsc.*, Stockholm.

SUITO, E., and UYEDA, N. (1957), *Proc. Imp. Acad. Japan* **33**, 398.

TERENIN, A. N., PUTZEIKO, E., and AKIMOV, I. (1957), *Zh. Fiz. Khim.* **31**, 716.

TERENIN, A. N., and SIDOROV, A. N. (1956), *Colloquium Spectroscopieum Internationale* **1**, Pergamon Press (London).

TOLLIN, G., KEARNS, D. R., and CALVIN, M. (1960), *J. Chem. Phys.* **32**, 1013.

TOLLIN, G., KEARNS, D. R., and CALVIN, M. (1961), *J. Chem. Phys.* **34**, 2022.

VARTANYAN, A. T. (1956), *Zh. Fiz. Khim.* p. 1028.

VARTANYAN, A. T., and KARPOVICH, I. A. (1958a), *Zh. Fiz. Khim.* **32**, 178.

VARTANYAN, A. T., and KARPOVICH, I. A. (1958b), *Zh. Fiz. Khim.* **32**, 274.

WHELAN, M. J., and HIRSCH, P. B. (1957a), *Phil. Mag.* **2**, 1121.

WHELAN, M. J., and HIRSCH, P. B. (1957b), *Phil. Mag.* **2**, 1303.

CHAPTER FIVE

Indirect Observation of the Bombarded Lattice

While direct observation of the crystal lattice during bombardment with fission fragments is clearly the ultimate goal in an investigation of this kind, it will be equally clear, from the observations made in the previous chapter, that we have at present a long way to go before achieving that end. Phase contrast electron microscopy imposes restrictive conditions on the nature and geometry of the crystal which is being 'imaged', there are severe difficulties associated with problems of interpretation of contrast patterns, and all experiment is conducted after, and not during irradiation. In this chapter, therefore, we begin to show how less direct methods may be employed in a study of the damage produced by an individual particle before proceeding, in Chapter 6, to a full and detailed discussion of the electron diffraction contrast at a fission particle track.

Image Formation for Thick Crystals

For crystal thicknesses greater than a few hundred Ångström units the approximations of the theoretical development in the previous chapter are no longer valid. Higher approximations described by Cowley and Moodie (1957), however, may be applied for limited thickness ranges. One immediate result is that approximations beyond the first order introduce amplitude as well as phase modulation due to pure elastic electron scattering. The theory becomes more complex.

No general solution to the problem of scattering from an arbitrarily thick crystal has yet been obtained, except in principle. It is normal therefore to make the assumption that there are only two strong beams in the crystal, the incident and one diffracted beam, and this is sometimes a good approximation for thick perfect crystals, suitably oriented. Cowley and Moodie (1957) derived the expression for the amplitude of the diffracted beam for this case, employing a new theoretical approach to the scattering of electrons by atoms and crystals, based on a new formulation of physical optics.† This becomes:

$$U_h = iF_h[((\sin wH)/w) \exp(i\sigma HE_0 - \rho HG_0 - \pi i \zeta_h H)] \qquad (5.1)$$

† In this formulation Huyghens' Principle is applied to give the amplitude distribution on one horizontal plane in terms of the amplitude distribution on preceding planes.

where $w = (\pi^2 \zeta_h^2 + F_h F_{\bar{h}})^{1/2}$, F_h being the complex structure factor given by $F_h = (\sigma E_h + i\rho G_h)$, ζ_h is the excitation error, or the distance in the beam direction of the Ewald sphere from the reciprocal lattice point denoted by h, and H is the crystal thickness.

Starting from the same general equation and considering only contributions to the amplitude in the incident beam direction, it is possible to derive in the same way the expression for the primary beam as:

$$U_0 = [\cos wH - i(\pi \zeta_h / w) \sin wH] \exp \{i\sigma H E_0 - \rho H G_0 + \pi i \zeta_h H\} \quad (5.2)$$

The intensity distribution at the exit face of the crystal is then given by:

$$\Psi\Psi^* = |U_0 + U_h \exp\{-2\pi i h x / a\}|^2$$

$$= 1 - \frac{2 \sin wH}{w} \left\{ \rho G_h \cos wH + \frac{\pi \zeta_h}{w} \sigma E_h \sin wH \right\} \cos 2\pi \left(\frac{hx}{a} - \zeta_h H \right)$$

$$- \frac{2 \sin wH}{w} \left\{ \sigma E_h \cos wH - \frac{\pi \zeta_h}{w} \rho G_h \sin wH \right\} \sin 2\pi \left(\frac{hx}{a} - \zeta_h H \right)$$

$$= 1 - 2|F_h| \frac{\sin wH}{w} \left(\cos^2 wH + \frac{\pi^2 \zeta_h^2}{w^2} \sin^2 wH \right)^{1/2} \cos 2\pi \left(\frac{hx}{a} - \zeta_h H - \alpha \right)$$

$$(5.3)$$

where

$$2\pi\alpha = \tan^{-1} \left\{ \frac{\sigma E_h \cos wH - (\pi \zeta_h / w) \rho G_h \sin wH}{\rho G_h \cos wH + (\pi \zeta_h / w) \sigma E_h \sin wH} \right\} \quad (5.4)$$

The image therefore consists of a set of cosine fringes which are representative of the lattice potential. In this case, however, unlike the circumstances discussed for *thin* crystals in Chapter 4, the amplitude of modulation, or contrast, and the phase depend in a complicated way on the modulus of the structure factor $|F_h|$, the excitation error ζ_h, and the crystal thickness H. Thus, when the crystal is perfectly disposed at the Bragg angle with respect to the incident electron beam in the microscope, equation 5.3 reduces to:

$$\Psi\Psi^* = 1 - \sin 2|F_h| H \cdot \cos 2\pi \left(\frac{hx}{a} - \alpha \right) \quad (5.5)$$

where $2\pi\alpha = \tan^{-1}((\sigma E_h)/(\rho G_h))$. The contrast of the lattice fringes therefore varies with crystal thickness. Furthermore, a full examination of equation 5.3 reveals anomalies in the direction and spacing of lattice fringes for images observed at or near to a Bragg extinction contour (Chadderton, 1961). It will be evident that it is necessary to exercise extreme caution in identifying fringe disturbances observed on *thick* crystals with lattice disturbances produced by irradiation.

Moiré Patterns

Yet there are ways in which we can exploit electron diffraction from thick crystals and use the electron microscope as a tool in the study of radiation damage. One of these methods requires the use of overlying crystals.

The successive scattering of electrons by two thin crystals separated by an arbitrary distance has been treated in detail by Cowley and Moodie (1959a) on the assumption that the scattering is kinematic. It was shown that, when certain relationships are satisfied between the dimensions and orientations of the projected potential distributions $\phi_1(xy)$ and $\phi_2(xy)$ and the separation of the two crystals, it is possible to obtain a periodic modulation of the intensity of the image with a periodicity *very much greater* than that of either crystal lattice. The intensity distribution is then given by:

$$\sum_h \sum_k E_1(h,k)\, E_2(-h,-k) \exp 2\pi i \left(\frac{hx}{A} + \frac{ky}{B}\right) = \varphi_1 \left(\frac{x}{M}, \frac{y}{M}\right) * \varphi_2 \left(-\frac{x}{M}, -\frac{y}{M}\right)$$

$$(5.6)$$

where A and B are the axes of the magnified unit-cell and M is the magnification. If the two crystals are identical, this reduces to the Patterson function for the crystal projection.

The results of the previous section allow the extension of this approach to the case where the scattering in the individual crystals is dynamic – where there is an amplitude as well as a phase effect. Since discussion is confined to dynamic scattering for the two-beam case only, it is convenient to deal with sets of parallel fringes with periodicity in one direction and so restrict the treatment to one dimension (Cowley and Moodie, 1959b). In the relevant direction the periodicities of the two crystals are considered to be a_1 and a_2, the structure factors are E_1 and E_2 (ignoring inelastic scattering contributions), the excitation errors are ζ_1 and ζ_2, and the crystal thicknesses are H_1 and H_2.

The wave incident on the second crystal is that at the exit face of the first crystal, namely:

$$\Psi_1(x) = U_0' + U_h' \exp(2\pi i h x / a_1) \tag{5.7}$$

Applying normal methods to calculate the scattering in the second crystal, it can be assumed that each term of the series which are summed to give U_0'' and U_h'' is therefore multiplied by $\Psi_1(x)$ instead of $\Psi_0(x) = 1$. At the exit face of the second crystal the intensity distribution is therefore given by:

$$\Psi\Psi^* = |\{U_0' + U_h' \exp(2\pi i h x / a_2)\}|^2$$

The sign of h has been changed for the second crystal since we must assume a reflection opposite to that of the first crystal in order to get a diffracted beam close to the primary beam. Evaluation of (5.8) gives:

$$\Psi\Psi^* = C_1^2 C_2^2 + \sigma^4 E_h^2 E_{\bar{h}}^2 \frac{\sin^2 w_1 H_1}{w_1^2} \cdot \frac{\sin^2 w_2 H_2}{w_2^2} - 2\sigma^2 E_h E_{\bar{h}} C_1 C_2$$

$$\times \frac{\sin w_1 H_1}{w_1} \cdot \frac{\sin w_2 H_2}{w_2} \cos 2\pi \left(\frac{hx}{A} - \zeta_1 H_1 - \zeta_2 H_2 + \beta_1 + \beta_2 \right) \qquad (5.9)$$

where

$$C_1 = \{ \cos^2 w_1 H_1 + (\pi^2 \zeta_1^2 / w_1^2) \sin^2 w_1 H_1 \}^{1/2} \qquad (5.10)$$

and

$$2\pi\beta_1 = \tan^{-1} \left\{ \frac{2\pi\zeta_1}{w_1} \tan w_1 H_1 \right\} = \frac{\pi}{2} - 2\pi\alpha_1 \qquad (5.11)$$

and similarly for C_2 and β_2, and $(1/A) = (1/a_1 - 1/a_2)$, A being the periodicity of the moiré-like fringes. In the derivation of (5.9), terms involving $\exp(2\pi i x/a_1)$ and $\exp(2\pi i x/a_2)$ have been omitted since these correspond to the spacings of the individual crystals – supposedly unresolved.

Hence the background intensity, the contrast, and the phase of the moiré fringes vary widely with the structure factors, crystal thicknesses, and excitation errors for the two crystals. Perhaps a better impression of the nature of the variations to be expected can be obtained by considering a special case. For example, if the crystals can be assumed parallel, we may take $\zeta_2 = -\zeta_1 = \zeta$. If one of the crystals, say the second, is very thin, we may take $\cos w H_2 = 1$, and $\sin w H_2 \approx w H_2$. Then, if we neglect terms in H_2^2, equation 5.9 reduces to:

$$\Psi\Psi^* = C_1^2 - 2H_2 \sigma^2 E_h E_{\bar{h}} C_1 \frac{\sin w_1 H_1}{w_1} \cos 2\pi \left\{ \frac{hx}{A} + 2\zeta H_1 + \beta_1 \right\} \qquad (5.12)$$

This equation is similar to equation 5.3 for a single thick crystal, $|F_h|$ being replaced by $H_2 \sigma^2 E_h E_{\bar{h}}$, the constant background being replaced by C_1^2 which depends on ζ_1, F_h, and H_1, and the phase angle α being replaced by $\beta = (\frac{1}{4} - \alpha)$. The moiré fringes therefore have much the same appearance as the fringes given by a single dynamically scattering crystal, and are subject to the same limitations of interpretation. One great difference, however, is the much greater separation of the moiré fringes owing to the effective magnification supplied by the composite double crystal – a separation which is well within the resolving power of modern electron microscopes. This single fact makes it possible indirectly to observe many crystal lattices which we are not yet directly able to resolve.

Diffraction Contrast

The first-order approximation of the Cowley-Moodie theory of the scattering of electrons by crystals provides a simple but very powerful means of interpreting electron microscope images of crystals. In fact nearly all of the observations which have been reported in the literature of contrast effects in the imaging of crystal lattices or their defects, whether the lattice is resolved or not, may be given an adequate qualitative and semi-quantitative explanation on this basis. The approximation, as applied to the study of periodic objects, is equally applicable when the periodicity is interrupted by lattice defects or the crystal is distorted. The simple phase grating concept allows an immediate appreciation of the nature of the modifications of the image involved and allows calculations to be made of the intensity distribution in particular cases.

In cases where the periodicity of the crystal lattice is not resolved, the intensity distribution due to a defect or distortion of the lattice may be calculated by using a development of an idea due to Doi (1957). It is assumed that the lattice periodicity is not resolved because all Bragg reflections are prevented from contributing to the image by limitation of the aperture or by aberration of the lenses. The effect is that of removing all of the diffraction pattern except for a small region surrounding the direct beam. An approximation to this effect is obtained by multiplying the diffraction pattern amplitude function by a function, say $g(xy)$, which is approximately unity within the effective aperture and very small outside the aperture, being zero at the Bragg reflection positions. Doi used a function of the type $g(x) = (\sin \pi ax)/\pi ax$, where a is the lattice periodicity, but other more realistic aperture functions may also be used (Cowley and Moodie, 1960).

The effect of multiplying the diffraction pattern by the aperture function $g(xy)$ is to convolute the image by its Fourier transform $G(xy)$:

$$\Psi(xy) = \Psi_0(xy) * G(xy) \tag{5.13}$$

Since $G(xy)$ will have a width greater than the unit-cell dimensions, the convolution will remove from the image all detail of the lattice periodicity, but slow variations in $\Psi_0(xy)$, such as are associated with radiation-induced lattice defects, will remain. They also appear in the expression for the intensity at the exit face of the crystal.

For thicker crystals a more satisfactory theory of diffraction contrast, a generalized dynamical theory, has been formulated by Howie and Whelan (1961). This is essentially a development of a two-beam dynamical approach from kinematical principles by methods which are similar to those used by Darwin (1914a, b) in a treatment of the Bragg case for X-rays. It permits an extension to the case of a deformed crystal and a derivation of the differential equations for direct and diffracted wave amplitudes in the crystal, from which the intensity is readily obtained. Fundamental to the theory is the so-called 'column approximation'. The distortions of columns of crystal which surround and lie at different distances from the core of the defective crystalline volume are considered in terms of a displacement vector **R** and the resulting contrast effects at and around the disturbances are computed from such considerations. In spite of strong objections to the 'column approximation' on what appear to be very sound grounds, there is immensely satisfying agreement between the theoretical images of dislocations and those which are observed experimentally. We shall see shortly that there is equally satisfying agreement for fission particle tracks, which exhibit structural features of contrast that are best interpreted in terms of the depth oscillations of the crystal wave fields in the dynamical region. Diffraction contrast, therefore, is another method which we can use as a tool of investigation in the physics of fission damage.

Experimental

Fission damage in two compounds of molybdenum was investigated. Crystals of molybdenum trioxide were prepared by direct condensation from the vapour onto the carbon support film of the electron microscope specimen grid. The smoke from burning molybdenum deposits crystals of the trioxide which are orthorhombic ($a_0 = 3 \cdot 962$ Å, $b_0 = 13 \cdot 858$ Å, $c_0 = 3 \cdot 697$ Å) and which lie on (010). Molybdenum disulphide specimens were obtained by successive cleavage with tweezers from the parent crystal and the resulting crystallites were attached to specimen supports by thin slivers of adhesive tape. It was discovered that the alternative method of repeated cleavage, with the use of the adhesive tape method suggested by Silk and Barnes (1959) and by Kamiya, Ando, Nonoyama, and Uyeda (1960), introduced a multiplicity of dislocations in the basal plane. Molybdenite has a hexagonal layer-like structure and cleaves parallel to (0001). The specimens were in general irradiated with randomly directed fission fragments in the manner which we have already described, though occasionally a collimated bombardment was employed.

Molybdenum Trioxide

Crystals of molybdenum trioxide irradiated with fission fragments and thermal neutrons possessed light contrast tracks and spots (Figure 24). Crystals irradiated with thermal neutrons only appeared unchanged. In molybdenum trioxide the width of tracks and the diameter of spots approached 160 Å while in platinum phthalocyanine (Chapter 4) the figure was nearer to 100 Å. Moreover, although the tracks and spots were again of light contrast on a dark background, they sometimes showed other new features. Thus, in Figure 25, the spots are accompanied by a surrounding dark contrast effect, and in Figure 26 the tracks have an intermittent nature.

When one crystal lay over another, and this frequently happened, the distortion of the lattice around the spots showed up very clearly in the moiré pattern.

Figure 24. Electron micrograph of fission fragment irradiated crystals of molybdenum trioxide. Spots (S) and tracks (T) are evident. Moiré patterns are formed in regions where crystals overlap.

It was often the case that one crystal lay on top of, *and* almost perpendicular to, another crystal. In these circumstances the moiré patterns were particularly sharp and clear; some such instances are shown in Figure 27. The tracks did not appear to distort the moiré pattern fringes very much. The fringes either stopped and started abruptly at the track or else continued across the track with reduced contrast.

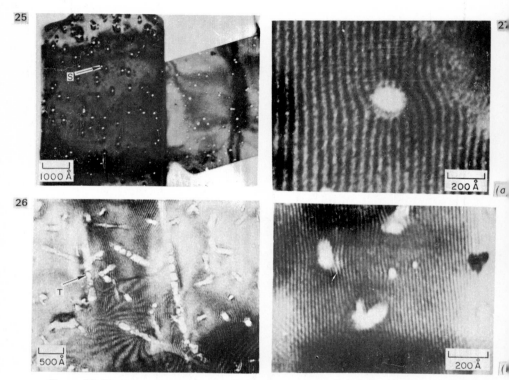

Figure 25. Electron micrograph of a fission fragment irradiated crystal of molybdenum trioxide. Regions of dark contrast surround the spots (S).

Figure 26. Electron micrograph of fission fragment irradiated crystals of molybdenum trioxide. Some of the tracks (T) are of intermittent character.

Figure 27. Electron micrographs of fission fragment irradiated crystals of molybdenum trioxide: (a) rotational moiré pattern; (b) parallel moiré pattern.

Molybdenum Disulphide

The contrast effects observed in fission fragment irradiated crystals of molybdenum disulphide were different from those in the trioxide. In the first instance the tracks and spots were of a dark contrast on a light background. A typical area is shown in Figure 28. No such tracks and spots were observed in crystals irradiated with thermal neutrons only. The tracks were sometimes intermittent

28

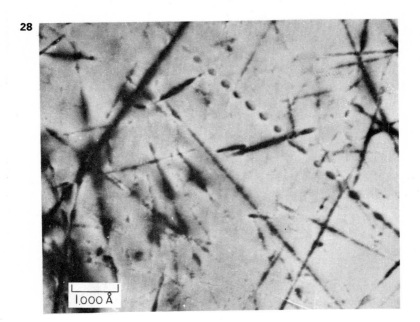

1,000 Å

29

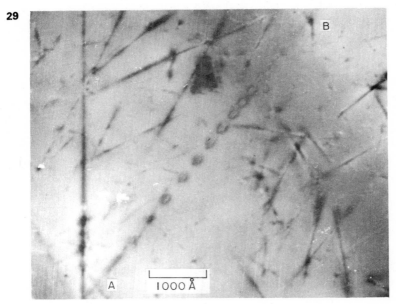

B

A   1000 Å

Figure 28. Electron micrograph of fission fragment irradiated crystal of molybdenum disulphide.

Figure 29. Electron micrograph of fission fragment irradiated crystal of molybdenum disulphide. The track AB is intermittent in character and consists of segments which are resolvable as dark contrast loops.

in nature and in certain circumstances, or when the specimen was tilted in the electron microscope, it proved possible to resolve the segments as dark contrast loops (Figure 29). In the neighbourhood of a Bragg extinction contour, the track contrast was observed to change and to become double in character (Figure 30), while, when lying immediately on an extinction contour, the tracks took on a black/white alternating contrast. Very occasionally the contrast loops moved under observation and particularly when the track was near one of the large dislocation arrays which are so often observed in molybdenum disulphide.

The track direction was often observed to change slightly but suddenly (Figure 31) and, in some cases, as in Figure 32, forked tracks were seen. Track interactions with dislocations already present were frequently observed (Figure 33), and moiré fringe patterns were perturbed (Figure 34).

Discussion

The tracks observed in thin crystals of molybdenum trioxide differ in nature from those obtained with crystals of platinum phthalocyanine. The comparatively large widths of some of the tracks indicate that the energy liberated by the moving particle in the lattice can cause considerably more damage than was the case with the phthalocyanines. The extreme light contrast of the tracks also suggests the removal or rearrangement of a large amount of crystalline matter. This is borne out in particular by the observation of holes some 160 Å in diameter which are hewn out of the crystal by fast-moving fragments. Molybdenum trioxide is comparatively stable in the electron microscope but can be decomposed to form small crystallites of molybdenum under the influence of electron beam currents. The considerable damage produced by a fission particle is therefore not surprising.

The intermittent nature (Figure 26) of some of the tracks is most interesting and there are two possible explanations. It may be that for some reason the energy of the moving fragment is communicated to the crystal lattice in distinctly separate bursts (see page 213, Chapter 13). Alternatively there may well be a rearrangement of the lattice damage resulting from a cylindrical pattern of disturbance and this may be such as to give the impression of an intermittent track. There is no evidence to suggest that dynamic electron diffraction effects at inclined cylindrical tracks would give rise to the type of light contrast observed.

The dark contrast effects which surround the light spots shown in Figure 25 may arise from a localized strain field which surrounds each spot. It has been shown in the previous chapter by direct observation of the lattice fringes that irradiated platinum phthalocyanine possesses such strain fields. It would not be surprising if the same occurred in molybdenum trioxide. In this respect the observations of the moiré pattern distortions shown in Figure 27 are particularly useful. The a_0 and c_0 lattice translations for orthorhombic molybdenum

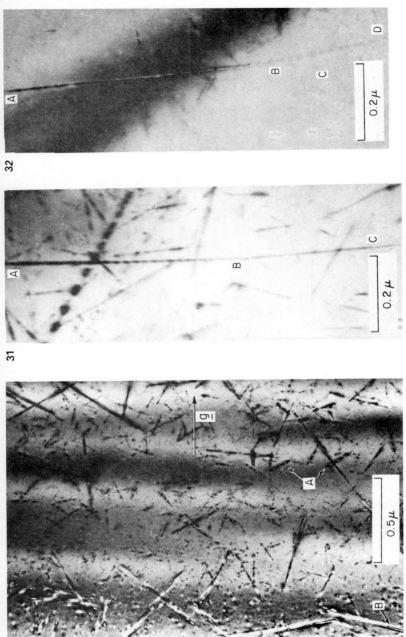

Figure 30. Electron micrographs of fission fragment irradiated crystal of molybdenum disulphide. The tracks vary in the type of contrast they show. Some have a double line contrast (A) and others have a black/white contrast (B).

Figure 31. Electron micrograph of fission fragment irradiated crystal of molybdenum disulphide. The track AC changes its direction at B.

Figure 32. Electron micrograph of fission fragment irradiated crystal of molybdenum disulphide. The track AD changes its direction at B and the struck atom registers as far as C.

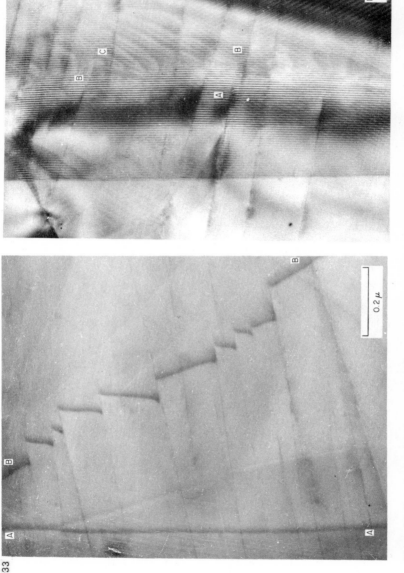

Figure 33. Rotation moiré pattern of (11$\bar{2}$1) planes in molybdenum disulphide. The fringe spacing, 40 Å, corresponds to a relative rotation of 2·2°, so the fringes are nearly perpendicular to the corresponding diffracting planes. The dislocation (B–B), revealed by diffraction contrast, has been 'cut' by intersecting tracks and the segments have relaxed to an orientation nearly parallel to the dislocation (A–A).

Figure 34. Damage revealed by disturbances of moiré patterns on molybdenum disulphide. The damaged region at A is about 70 Å wide. The fringe patterns at B and C are discontinuous across the tracks; extra half-fringes have appeared at B.

trioxide are very nearly the same, so that, in preparing the crystals from the vapour, epitaxy often leads to one crystal growing on top of, and at right-angles to, another. In these circumstances, double diffraction will lead to *parallel* type moiré patterns and such patterns may, as we have seen, reveal the presence of defects in one or both of the crystals involved. Suppose it is assumed, as was the case for platinum phthalocyanine, that the molecular planes in both crystals of molybdenum trioxide curve inward toward the region of damage. Interference between the (001) planes of the first crystal and the (100) planes of the second crystal then leads to a moiré pattern of the type shown

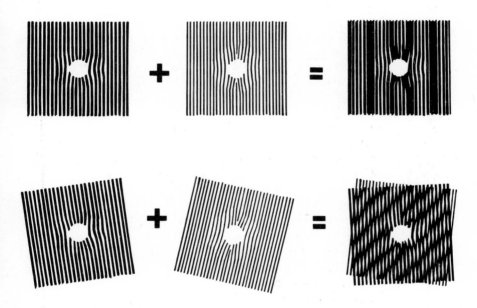

Figure 35. Optical analogue showing the appearance of damaged regions in moiré patterns: (a) parallel moiré pattern; (b) rotational moiré pattern.

in Figure 35(a). This should be compared with Figure 27. The assumption of inward curving molecular planes appears to be proved. In the event of a misorientation of the two crystals, *rotational* moiré patterns occur, and an example is shown in Figure 35(b). Such features have also been seen in practice.

It may again be assumed that, when in the thicker crystals the moiré patterns simply cross the light contrast tracks, then this implies that the perfect crystal above and below the track swamps the distortion due to the cylindrical region of damage.

The observation of dark tracks which change their contrast on tilting the specimen suggests that these are essentially due to diffraction contrast. Variations in the Bragg conditions occur at the tracks, which probably have a crystallographic orientation somewhat different from the undamaged crystal. Many observers have seen similar contrast variations at inclined dislocations in metal foils and there are other parallel contrast phenomena also. In particular, the double line contrast shown in Figure 30 is a feature which is familiar to defect solid-state electron microscopists. It occurs when two sets of reflecting planes operate simultaneously on opposite sides of the disturbance. Such a doubling of contrast in the image of a dislocation is able to be removed by tilting the specimen *in situ*. This is also observed with dark contrast tracks in the irradiated specimens of molybdenum disulphide. The black/white contrast, common to both tracks and dislocations, can also be removed by tilting the specimen. Two points should be emphasized. Firstly, the similarities between fission fragment disturbances and dislocations must not be interpreted as necessarily implying equivalence. Secondly, the periodic contrast variations, whether they are of a black/white or intermittent nature, are not likely to be structural features of the tracks.

There is, however, distinct evidence that dislocations are present in some parts of the fission fragment tracks. It is undoubtedly true that some of the intermittent tracks possess this property only because of their inclination to the plane of the crystal (see the next chapter). In other circumstances, however, the segments of a particular track do not have a constant separation and sometimes eventually merge to form a continuous track. This indicates the presence of a real structure in the tracks. Identical features have been observed by Silk and Barnes (1959) in irradiated mica. The observation that the segments can be resolved as loops of dark contrast with diameters up to 200 Å may be interpreted in two ways. Either the contrast arises from a surrounding strain field (as with molybdenum trioxide) or else it is directly due to a closed dislocation loop. The latter possibility certainly occurs on occasions, since the loops are sometimes seen to move in the neighbourhood of dislocations present in the crystal before irradiation. Ruedl, Delavignette, and Amelinckx (1961) attribute similar movements of dislocation loops in platinum to conservative climb of the kind described by Kroupa and Price (1961) for zinc. The further possibility that a partial annealing of the tracks owing to the heat arising from electron irradiation in the microscope may lead to tracks with a dashed nature is discounted. When irradiated specimens are deliberately subjected to high beam currents there is no apparent change in the number or nature of tracks.

The sudden changes in track direction of the kind shown in Figure 31 are probably a result of nuclear encounters of the fission fragments with lattice atoms which result in a scattering of the fragments in the plane of the crystal. In the few cases where forked tracks are observed the struck atoms also produce a track. In these cases both the struck atom and the fission fragment must have sufficient energy after the event to register as tracks in the crystal and this implies a considerable transfer of energy to the lattice atom. Such an event is shown in Figure 32 and a detailed analysis of this electron micrograph performed on the basis of that given by Noggle and Stiegler (1960) indicates that the knock-on is a sulphur atom.

The presence of disjointed dislocations at intersections with fission particle tracks (Figure 33) probably implies a 'jogging' of the moving dislocation line by the disturbance produced by a fission fragment or a mutual interaction of the two strain fields. In many cases the dislocation lines pass over the tracks, so it is reasonable to assume that the damage along the length of the track is localized over a region about 100 Å in diameter; though no very accurate estimation of the diameter is possible from the observations since diffraction contrast is the mechanism which is operating.

Conclusion

Electron microscopy of fission fragment irradiated molybdenum trioxide and molybdenum disulphide indicates that individual fission particle tracks may be imaged by either absorption contrast (light tracks) or diffraction contrast (dark tracks). Certain features in the electron micrographs are undoubtedly contrast artefacts and are not real in nature. Other features, however, are most certainly real and are indicative of the way in which the energy of the moving particle is converted into crystal defects. They merit further more detailed investigation. Moiré patterns on overlying crystals may be used indirectly to 'see' the damaged lattice.

References

CHADDERTON, L. T. (1961), *Nature*, Lond. **189**, 564.

COWLEY, J. M., and MOODIE, A. F. (1957), *Acta Cryst.* **10**, 609.

COWLEY, J. M., and MOODIE, A. F. (1959a), *Acta Cryst.* **12**, 423.

COWLEY, J. M., and MOODIE, A. F. (1959b), *Acta Cryst.* **12**, 367.

COWLEY, J. M., and MOODIE, A. F. (1960), *Proc. Europ. Reg. Conf. Electr. Microsc.*, Delft, **I**, 199.

DARWIN, C. G. (1914a), *Phil. Mag.* **27**, 315

DARWIN, C. G. (1914b), *Phil. Mag.* **27**, 675.

DOI, K. (1957), *Bull. Soc. Franc. Miner. Crist.* **80**, 325.

HOWIE, A., and WHELAN, M. J. (1961), *Proc. Roy. Soc. A* **263**, 217.

KAMIYA, Y., ANDO, K., NONOYAMA, M., and UYEDA, R. (1960), *J. Phys. Soc. Japan* **15**, 2025.

KROUPA, F., and PRICE, P. B. (1961), *Phil. Mag.* **6**, 243.

NOGGLE, T. S., and STIEGLER, J. O. (1960), *J. Appl. Phys.* **31**, 2199.

RUEDL, E., DELAVIGNETTE, P., and AMELINCKX, S. (1961), *J. Appl. Phys.* **32**, 2402.

SILK, E. C. H., and BARNES, R. S. (1959), *Phil. Mag.* **4**, 970.

CHAPTER SIX

The Imaging of Fission Fragment Tracks by Electron Diffraction Contrast

The discussion of electron diffraction contrast which we gave in Chapter 4 was developed on the assumption that in the electron microscope the crystal behaves like a phase grating. In this approximation, which may be applied most satisfactorily to the thinnest crystals (approximately 100 Å), the wave functions for the electron waves at the entry and exit surfaces of the crystal are related in a particularly simple way. For thicker crystals, however, there is an amplitude grating behaviour and it is necessary to consider in detail what happens to the electron waves *inside* the crystal, where complicated absorption and scattering processes are taking place.

We have also seen that the nature of the fission fragment damage which may be observed in the electron microscope is not only dependent on the nature of the irradiated material but changes with specimen thickness. In the thinner crystals each track, if present, is generally revealed as a continuous or inter- mittent light line on a dark ground that is evidently a region of enhanced elec- tron transmission. For thicker crystals the tracks are dark on a light ground. It may therefore be concluded that for a light contrast track part of the crystal is *removed* by the particle so that a track is revealed by 'absorption contrast'. In thicker crystals, not all of the material is removed along the path of the fragment, though atoms are evidently *rearranged* and strains are established in the lattice. This results in localized perturbations of the prevailing electron diffraction conditions and 'diffraction contrast' (Hirsch, Howie, and Whelan, 1960) begins to operate. In this chapter a model will be presented for the damage produced by a fission fragment in the thicker crystals (approximately 1000 Å), and the two-beam dynamical theory of electron diffraction contrast (Howie and Whelan, 1961) will be used to investigate the nature of the contrast profile of a track for different circumstances (Chadderton, 1964). A discussion on the basis of the kinematical theory of diffraction (Hirsch *et al.*, 1960) may be shown to lead to qualitatively similar results (Chadderton, 1963).

The Model

The model which we choose is in fact based on direct observations of the bombarded lattice described in Chapter 4 and is in many respects the cylindrical equivalent of the spherically symmetrical coherency strain of Ashby and

Brown (1963a). In the case of a spherical precipitate particle embedded, with misfit, in an infinite isotropic matrix, the displacements are radial, so that the displacement function $\mathbf{u}_r$ (in the matrix) may be given in terms of the distance r from the centre of the particle by:

$$\mathbf{u}_r = \epsilon r_0^3 \mathbf{r}/r^3 \quad (r \geqslant r_0) \tag{6.1}$$

where r_0 is the radius of the precipitate inclusion and ϵ is related to δ, the misfit between the unstrained lattices of inclusion and matrix (Mott and Nabarro, 1940). Consequently, for a fission fragment track, a central cylinder of damage of radius r_0 is assumed to be initiated by the particle, though the

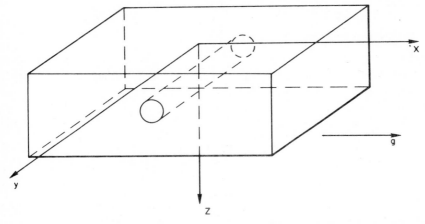

Figure 36. Co-ordinate system used to indicate the position of a fission fragment track within the crystal. The reflection vector **g** is directed along the positive direction of the x-axis.

structure of the contents of this cylinder is not defined at this time. Beyond the bounds of the cylinder is a strain field for which, as in the case of the precipitate particle, the displacements are radial. In order to obtain a relation for the corresponding displacement function $\mathbf{u}_r$, the medium is assumed to be elastically isotropic and only a thin crystal slice cut perpendicular to the track, which is assumed horizontal in the specimen (Figure 36), is considered. Now, following a treatment by Love (1927), the components of the displacements in plane strain, corresponding to a single force of magnitude $2\pi A$ acting in the negative direction of the x-axis at the origin, may be shown to be given by:

$$u_x = \frac{(\lambda + 3\mu)}{2\mu(\lambda + 2\mu)} \, A \log_e r + \frac{(\lambda + \mu) \, A y^2}{2\mu(\lambda + 2\mu) r^2} \tag{6.2}$$

$$u_y = -\frac{(\lambda + \mu) \, A x y}{2\mu(\lambda + 2\mu)} \tag{6.3}$$

where λ and μ are Lame's (1852) coefficients. We note that λ and μ are related to the elastic moduli in the following fashion:

$$\text{modulus of compression, } k = \lambda + (2\mu/3) \tag{6.4}$$

$$\text{Young's modulus, } E = \frac{\mu(3\lambda + 2\mu)}{\lambda + \mu} \tag{6.5}$$

$$\text{Poisson's ratio, } \sigma = \frac{\lambda}{2(\lambda + \mu)} \tag{6.6}$$

and that μ itself is often referred to as the 'rigidity'. For our purposes, however, a more useful type of singular point may be derived by supposing that the following forces are applied near the origin:

(a) parallel to the x-axis $- 2\pi A$ at the origin and $2\pi A$ at $(h,0)$
(b) parallel to the y-axis $- 2\pi A$ at the origin and $2\pi A$ at $(0,h)$

We may now pass to a limit by supposing that Ah remains constantly equal to $\pm B$ while h is diminished without limit. The components of the displacements in plane strain are then given as follows:

$$u_x = \frac{\pm B(\lambda + 3\mu)}{2\mu(\lambda + 2\mu)} \frac{\partial}{\partial x} (\log_e r) + \frac{\pm B(\lambda + \mu)}{2\mu(\lambda + 2\mu)} \left(\frac{\partial}{\partial x} \frac{y^2}{r^2} - \frac{\partial}{\partial y} \frac{xy}{r^2} \right) \tag{6.7}$$

$$u_y = \frac{\pm B(\lambda + 3\mu)}{2\mu(\lambda + 2\mu)} \frac{\partial}{\partial y} (\log_e r) + \frac{\pm B(\lambda + \mu)}{2\mu(\lambda + 2\mu)} \left(\frac{\partial}{\partial y} \frac{x^2}{r^2} - \frac{\partial}{\partial x} \frac{xy}{r^2} \right) \tag{6.8}$$

or by:

$$(u_x, u_y) = \frac{\pm B}{\lambda + 2\mu} \left(\frac{\partial}{\partial x}, \frac{\partial}{\partial y} \right) \log_e r \tag{6.9}$$

In polar co-ordinates we may write:

$$\mathbf{u}_r = \frac{\pm B}{\lambda + 2\mu} \frac{\mathbf{r}}{r^2} \tag{6.10}$$

$$\mathbf{u}_\theta = 0 \tag{6.11}$$

This strain field is particularly suitable for our purposes since both its strength and nature may be adjusted by altering the sign and magnitude of B. It is therefore a simple matter to construct a centre of compression or contraction. Moreover, the logarithmic divergent nature of the function is simply avoided by placing the origin in a circular cavity of radius r_0. This is just what we require. A uniform pressure of amount:

$$\frac{\pm 2\mu B}{(\lambda + 2\mu)} \frac{1}{r_0^2} \tag{6.12}$$

is exerted on the walls of the cavity. Equations 6.10 and 6.11 apply equally well, of course, to the cylindrical case, in which a large number of thin slices are laid together. There is, however, the additional condition that displacements along the direction of the axis of the track (Figure 36) are assumed negligible:

$$\mathbf{u}_y = 0 \quad (r \geqslant r_0) \tag{6.13}$$

Equation 6.10, which predicts the displacement of atoms at a distance r from the centre of the 'core' of the track, should be compared with equation 6.1 for the spherically symmetric coherency strain. It will be clear that, as a result of adopting a model with cylindrical symmetry, the strain field 'reaches farther

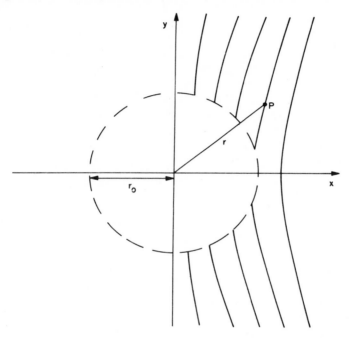

Figure 37. Diagrammatic representation of the model used to simulate the damage produced by a fission fragment in a crystal lattice. A radial strain field surrounds a cylinder of radius r_0. The contents of the latter are not defined.

out' into the surrounding crystal and that, at a given distance r for a given strength of strain field, the displacement of an atom is greater than was the case for the precipitate particle. In what follows, the negative sign will be adopted in equation 6.10 so that the central cylinder of radius r_0 will be assumed to act as centre of contraction for the surrounding crystal (Figure 37).

Theory and Results

A full treatment of diffraction contrast effects in the electron microscope has been given by Hirsch *et al.* (1960) and by Howie and Whelan (1961). In the kinematical theory (Hirsch *et al.*, 1960) the contrast profile at a defect, such as a screw dislocation, is obtained from a simple amplitude/phase diagram construction. Some features of the contrast effects which are observed in practice may be explained in this manner, but in general it is necessary to take account of absorption, the existence of several diffracted beams, and a dynamical interaction between incident and diffracted beams. For this reason the generalized dynamical theory (Howie and Whelan, 1961) has been established, and this is much more successful in explaining the complicated nature of contrast profiles at defects such as dislocations (Howie and Whelan, 1962), stacking faults (Hashimoto, Howie, and Whelan, 1962), and precipitate particles (Ashby and Brown, 1963a, b). In this section the dynamical theory will be applied to the model which we have chosen to represent a fission fragment track (equation 6.10) and, in order to make an assessment of the validity of the model and its usefulness in radiation damage studies, the theoretical conclusions will be compared with experimental results.

Consider first of all the kinematical theory, and the co-ordinate system of Figure 36 with the track disposed horizontally in the centre of the crystal. The reflection vector $\mathbf{g}$ for the one strong diffracted beam is assumed to be directed along the x-axis. The amplitude of a single diffracted beam, neglecting absorption (Howie and Whelan 1961), is then essentially given by:

$$A_g = \frac{i\pi}{\xi_g} A_0 \int_0^t \exp\left(-2\pi i(s_g z + \mathbf{g} \cdot \mathbf{u})\right) dz \qquad (6.14)$$

where t is the crystal thickness, A_0 is the amplitude of the transmitted beam, ξ_g is the extinction distance corresponding to the reflection vector $\mathbf{g}$, and s_g indicates the extent of the deviation from the Bragg angle (the 'excitation error' of the Bethe (1928) terminology). We may now substitute for $\mathbf{u}_r$ from equation 6.10. Then, since only displacements perpendicular to the reflecting planes will produce contrast:

$$A_g = \frac{i\pi}{\xi_g} A_0 \int_0^t \exp\left[-2\pi i \left(s_g z - |g| \frac{Bx}{(\lambda + 2\mu)(x^2 + (v - z)^2)} \right) \right] dz \quad (6.15)$$

and this expression gives the amplitude diffracted by a column of crystal distant x from the track. The integral itself may be considered to define an amplitude/phase diagram, where z is the distance measured *along* the curve and where the slope at any point may be derived from the exponent. For a perfect crystal this amplitude/phase diagram is a circle of radius $(2\pi s_g)^{-1}$. However, when a particle track is introduced the crystal becomes imperfect in the way we have discussed, so that, instead of a circle, a spiral is generated. Thus, suppose that we define the position of a column of crystal, distant x from the track, through the parameter $\beta = 2\pi s_g x = +1$. Then the integral down the column as far as the track itself defines a coiled-up spiral, while the integral from the track to the bottom of the crystal defines an unwound spiral. Exactly the reverse occurs if another column of crystal equidistant from the track, but on the other side ($\beta = -1$), is considered. We may therefore conclude that the bright-field image will be symmetrical for this particular case. Moreover, since for $x = 0$, equation 6.15 reduces to the expression for a perfect crystal – as was discovered for the spherically symmetrical coherency strain (Ashby and Brown, 1963a) – the image of the track must be characterized by a central line of no contrast. By proceeding in this fashion, contrast profiles for tracks at quite general positions in the crystal may be obtained. The normal procedure is to construct graphically a series of amplitude/phase diagrams for a track in a given position, or, alternatively, to evaluate numerically the integral in equation 6.15. The loci of the centres of the initial and final circles on the amplitude/phase diagram are also easily obtained by numerical computation.

One further important observation results from the kinematical treatment. Suppose that the track is inclined with respect to the surface of the specimen, as might be expected if the fragment entered the crystal at the top surface and left it at the bottom. Segments of the track which lie at different depths in the crystal can now give rise to different diffracted amplitude vectors. In fact, if we move our column of reference through the crystal along a line parallel with and equidistant from the track itself, it may be demonstrated that the diffracted amplitude vector oscillates between two extremes. The track will therefore *appear* to be intermittent.

Consider now the same problem on the basis of the dynamical theory of electron diffraction contrast (Howie and Whelan, 1961). In the two-beam dynamical theory A_0 and A_g, the amplitudes of the transmitted and diffracted beams respectively, are given by:

$$\frac{\mathrm{d}A_0}{\mathrm{d}z} = -\frac{\pi}{\xi'} A_0 + \pi \left(\frac{\mathrm{i}}{\xi_g} - \frac{1}{\xi_g'} \right) A_g \qquad (6.16)$$

$$\frac{\mathrm{d}A_g}{\mathrm{d}z} = \pi \left(\frac{\mathrm{i}}{\xi_g} - \frac{1}{\xi_g'} \right) A_0 + \left\{ -\frac{\pi}{\xi_0'} + 2\pi\mathrm{i} \left[s_g + \frac{\mathrm{d}}{\mathrm{d}z} (\mathbf{g} \cdot \mathbf{u}) \right] \right\} A_g \qquad (6.17)$$

so that $|A_0|^2$ and $|A_g|^2$ define the transmitted and diffracted intensities respectively. ξ_0' and ξ_g' are parameters which are introduced by making the normal extinction distances complex:

$$\frac{1}{\xi_0} \rightarrow \frac{1}{\xi_0} + \frac{\mathrm{i}}{\xi_0'} \quad \text{and} \quad \frac{1}{\xi_g} \rightarrow \frac{1}{\xi_g} + \frac{\mathrm{i}}{\xi_g'} \qquad (6.18)$$

so that absorption of electrons (i.e. inelastic scattering) might be accounted for. Equations 6.16 and 6.17 have been numerically integrated, for a crystal containing a fission fragment track parallel to its surface, by the Runge-Kutta-Gill method of the high-speed digital computer, EDSAC 2. As in the case of the computation reported by Ashby and Brown (1963a), dimensionless variables

$(W = s_g \xi_g, X = x/\xi_g, V = v/\xi_g, T = \tau/\xi_g, R_0 = r_0/\xi_g)$ have been used throughout. In order to facilitate comparison with the results of Ashby and Brown, only the contrast profiles of the (111) copper reflection ($\xi_g = 222$ Å) have been examined extensively. Nevertheless the results may be considered to be generally applicable to any crystal system for which the two-beam theory and the column approximation are valid. The parameter which describes the strength of the cylinder of influence of the track is $gB/\xi_g(\lambda + 2\mu)$. This should be compared with the factor $\epsilon g r_0^3/\xi_g^2$ for the spherically symmetrical coherency strain, and with the quantity $\mathbf{g} \cdot \mathbf{b}$ for dislocations.

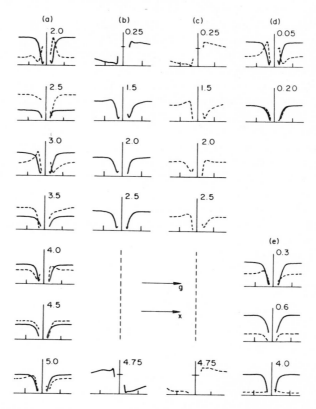

Figure 38. Diagram showing the dependence on different parameters of the image profile for a horizontal track. (a) As a function of crystal thickness where the track is centrally located ($V = T/2$, $R_0 = 0.25$, $|gB/\xi_g(\lambda + 2\mu)| = 0.159$, $s_g = 0$, $\xi_g/\xi_g' = \xi_g/\xi_0' = 0.1$) and for both bright (full line) and dark field (broken line). Values of T are indicated. (b) Bright-field profiles as a function of depth of the track below the top surface of the crystal ($T = 5.0$, $R_0 = 0.25$, $s_g = 0$, $|gB/\xi_g (\lambda + 2\mu)| = 0.159$, $\xi_g/\xi_g' = \xi_g/\xi_0' = 0.1$). Values of V are indicated. (c) Same but for dark field. (d) As a function of the absorption parameter. ($T = 5.0$, $V = T/2$, $R_0 = 0.25$, $|gB/\xi_g(\lambda + 2\mu)| = 0.159$, $s_g = 0$). Values of $\xi_g/\xi_g' = \xi_g/\xi_0'$ are indicated. (e) As a function of deviation from the Bragg condition ($T = 5.0$, $V = T/2$, $R_0 = 0.25$, $|gB/\xi_g(\lambda + 2\mu)| = 0.159$, $\xi_g/\xi_g' = \xi_g/\xi_0' = 0.1$). Values of s_g are indicated.

The Contrast Profiles

The results of some of the machine calculations are plotted as contrast profiles in Figure 38. The profiles have been obtained by moving the column of reference in the crystal through the track in the direction of the x-axis, beginning in relatively perfect crystal on one side of the track and ending at a point equidistant from the axis of the track but on the opposite side. The curves are cut off sharply at the surface of the inner cylinder ($R_0 = 0.25$), whose contents have not been defined. All the curves correspond to a dilatational strain field for which $|gB/\xi_g(\lambda + 2\mu)| = 0.159$.

In Figure 38(a), the contrast profiles in bright field (full line) and dark field (broken lines) are presented as a function of the crystal thickness, where the track is assumed to be horizontal and central in the specimen ($V = T/2$) and where the Bragg condition is perfectly satisfied ($W = 0$). In each case, ξ_g/ξ_g' and ξ_g/ξ_0' are taken to be 0.1. The bright-field images are perfectly symmetrical, and only the relative intensities change with thickness, bringing the bright- and dark-field images more nearly the same with increasing thickness. This is to be expected (Hashimoto *et al.*, 1962). The dark-field images on the other hand are markedly asymmetric, though they become less so as the crystal becomes thicker. Peaks and troughs in the dark-field profile oscillate from one side of the track to the other with a period of one extinction distance (ξ_g).

Figure 38(b) illustrates the way in which bright-field contrast profiles are expected to vary as the depth of the track in the crystal is changed. It is clear that the symmetry is only to be expected when the track is centrally located. As the track approaches one of the two free surfaces, however, the influence of 'image' forces becomes more effective, so that a pronounced asymmetry results. While the asymmetry is in opposite directions for the bright-field image where the track is considered to lie in two positions successively (first near the top and then near the bottom of the crystal), for the dark-field image (Figure 38(c)) it is in both cases in the same direction. We shall see later that this fact may be used with profit in radiation damage investigations.

Contrast profiles showing the variations which are to be expected if different absorption parameters ($\xi_g/\xi_g' = \xi_g/\xi_0' = 0.05$ or 0.2) are utilized in the calculations are given in Figure 38(d). In neither case is the agreement with experiment (see below) at all satisfactory; the dark-field image profile is either too asymmetric for a centrally located track, or alternatively it is too much like the corresponding image in bright field. It is therefore suggested that the best value for ξ_g/ξ_g' is of the order of 0.1, and all other profiles in Figure 38 are computed on this basis. The profile variations which occur with deviation from the Bragg condition are presented in Figure 38(e). The track is again horizontal and centrally disposed. The symmetry of the bright-field image is preserved but, for $s_g = 1$, the dark-field image has almost disappeared. This is not the case if s_g is varied for a track located at a more general position in the crystal, such as one near to the top or bottom surfaces. Then, for $s_g = 1$, the asymmetry of both bright- and dark-field images (Figure 38(b) and (c)) is maintained.

Comparison with Experiment

While some of the details of the profiles in Figure 38 are not observed in practice, the general experimental picture may be considered to be one of broad agreement. On the basis of the simple theory which has been outlined above, one would not expect tracks which are aligned in the direction of the reflection vector **g** to be revealed, or that, if present, the contrast would be weak. This may be confirmed by experiment (Figure 30), and the conditions correspond to those for which the extinction criterion for dislocations ($\mathbf{g.b} = 0$) is valid. Moreover, for tracks in the centre of the irradiated crystal which are inclined at a very low angle to the parallel surfaces, a central line of no contrast is indeed observed (Figure 39), and the contrast profile as revealed by a microdensitometer trace shows remarkably good agreement with that predicted by the theory. The agreement for dark-field profiles is not so accurate but is qualitatively satisfactory. Double images of tracks are also observed when it is clear that s_g is not equal to zero. In these cases (at D in Figure 30, for example) it may often be demonstrated that two reflection vectors g_1 and g_2 were excited simultaneously, so that the contrast variations across a track are an amalgam, as it were, of two separate profiles. For one reflection only, however, and for $s_g = 0$, *one* dark line appears to be dominant in the bright-field image. This is less satisfactorily explained on the basis of the theory, though for general positions of the track both bright- and dark-field image profiles do become asymmetric. In such circumstances if a 'band of dark contrast' is swept across the field of view by gradually tilting the crystal with respect to the direction of the electron beam then the sign of the operating reflection vector g can be reversed, and the side of the track on which the strong dark line of the image lies can be changed (Figure 40). It is possible, however, that this effect may only be observed when the track is near to a crystal surface, so that the image becomes anomalous.

Intensity oscillations with depth are frequently observed on inclined particle tracks in both bright and dark field, providing that the extinction distance ξ_g is low enough to permit a dynamic interplay between incident and diffracted beams for a crystal of given thickness t. Examples of this effect are given in Figure 30 (marked A) and Figure 41, where there is a doubling of the images and a central line of no contrast. The micrograph in Figure 41 shows tracks produced by a collimated beam of fission fragments (Bowden and Caspar, 1963) in molybdenum disulphide ($\xi_g = \xi_{110} = 504$ Å). A remarkable agreement between experiment and theory may be demonstrated for the case of oscillatory contrast effects on tracks It is possible, of course, to construct an intensity contour map for a track inclined at a given angle in a crystal of given thickness t and extinction distance ξ_g. In general, however, this is an extremely tedious computation, so we are content to show some experimentally obtained intensity profiles for comparison with those in Figure 38(b). In Figure 42, therefore, is a track in nickel chloride ($\xi_g = \xi_{110} = 185$ Å) which was observed and photographed in bright field. At the points marked a microdensitometer has been

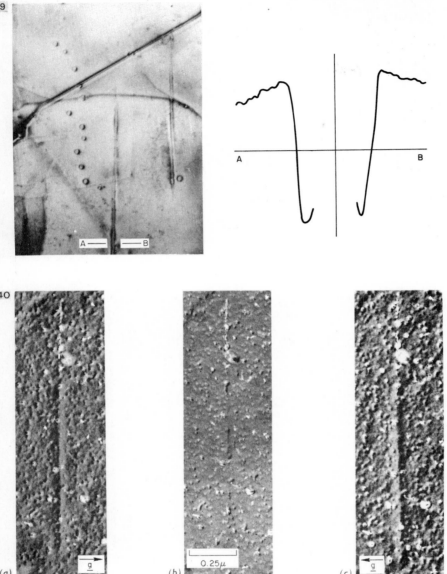

Figure 39. Electron micrograph showing fission fragment tracks in molybdenum disulphide. Only one reflection vector operates. The diagram is of a microdensitometer trace across AB. *Figure 40*. Electron micrographs showing the variation in contrast at a fission fragment track in lead iodide (PbI_2) when the crystal is tilted in the electron microscope so that a 'band of dark contrast' crosses the field of view. On reversing the sign of g a line of dark contrast in the image appears on the opposite side of the track itself.

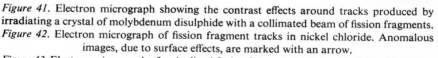

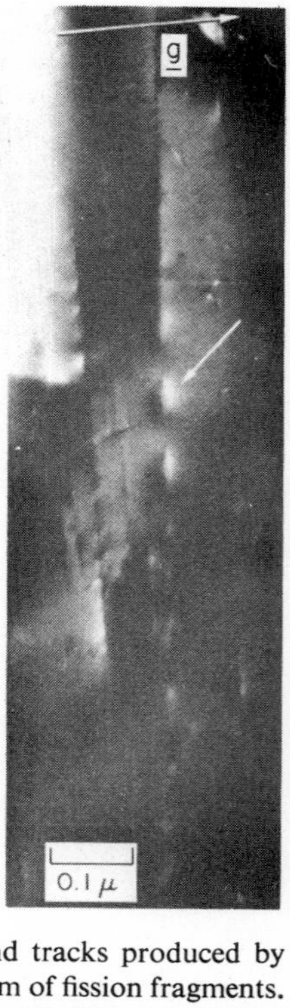

Figure 41. Electron micrograph showing the contrast effects around tracks produced by irradiating a crystal of molybdenum disulphide with a collimated beam of fission fragments.

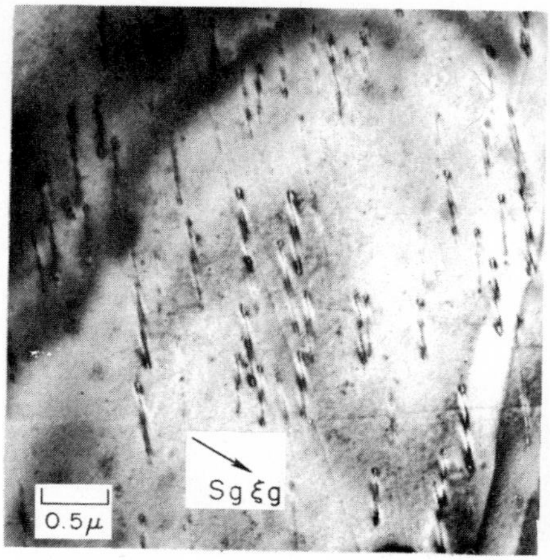

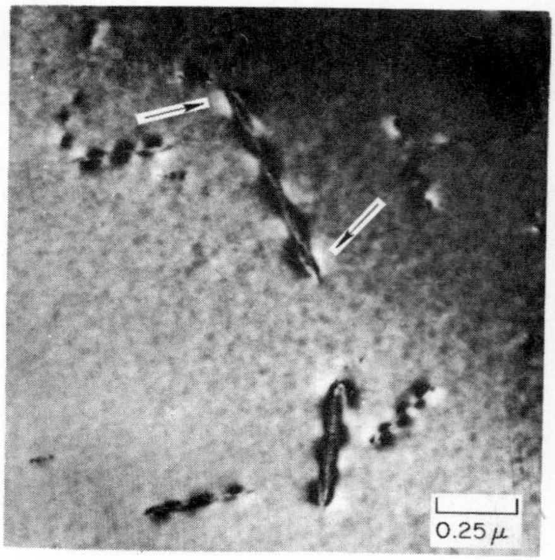

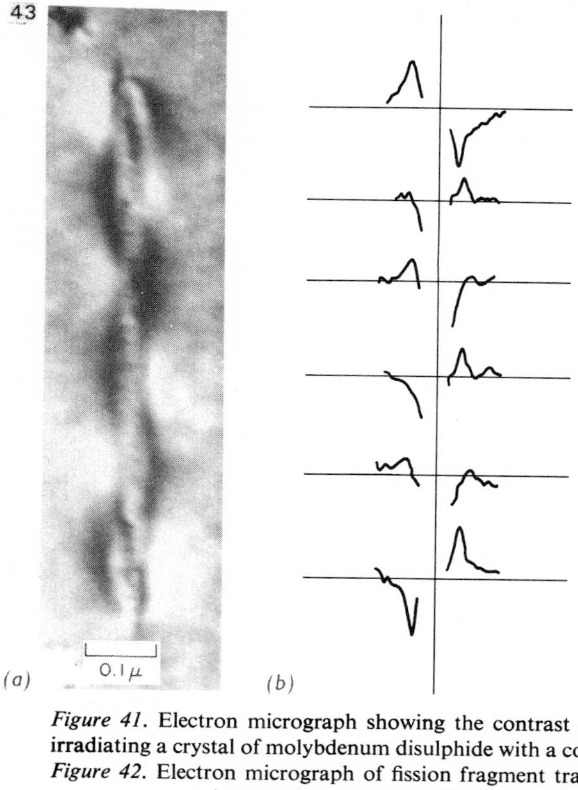

Figure 42. Electron micrograph of fission fragment tracks in nickel chloride. Anomalous images, due to surface effects, are marked with an arrow.

Figure 43. Electron micrograph of an inclined fission fragment track in nickel chloride. There is a central cylinder of 'no contrast', and oscillatory contrast effects are visible. Densitometer traces across appropriate parts of the track are shown in the adjoining diagram.

Figure 44. Asymmetric dark-field image of fission fragment track in molybdenum disulphide. The direction of the asymmetry indicates the existence of a centre of dilatation.

scanned across the track, and the profiles thus obtained have been plotted out for comparison with Figure 38(b). The qualitative agreement is very satisfactory, and anomalous contrast effects at the surfaces are clearly evident. Other features which are interpreted as being due to anomalous contrast are arrowed in Figure 42.

Discussion

The measure of agreement between theory and practice clearly suggests that the simple model of equation 6.10 might to a large degree describe the nature of the strain about a continuous fission fragment track. It should be remembered, however, that while the column approximation is a good one the other criterion, that of two beams only, is not so easily satisfied for inorganic covalent and ionic compounds (in which tracks are most frequently observed) as it is in metals. Nevertheless it is usually possible to photograph tracks at, and in the neighbourhood of a single strong extinction contour. Hence on these micrographs, if the model is a good one, there should at least be qualitative agreement between experiment and theory. It may therefore be concluded that the model of equation 6.10 is one, though admittedly perhaps not the only one, which satisfactorily explains the experimental results.

From the point of view of the radiation damage studies there are a number of further conclusions which may be made. We have noted, for example, that not all tracks revealed by diffraction contrast are continuous (Bowden and Chadderton, 1962) and that sometimes tracks must be discontinuous because of the mechanism of energy loss (page 27, Chapter 2). Since the separation between the segments of a discontinuous track is often of the order of a few hundred Ångströms ($\sim \xi_g$) it is clear that proper precautions must be taken to ensure that the intermittent nature is more than apparent. The proper contrast experiments should be performed by making use of the tilting device in the electron microscope.

The observation of anomalous contrast effects where a track meets the crystal surface suggests an interesting possbility. Thus we have observed that while the asymmetry of the bright-field image is different for tracks near to the top and bottom surfaces this is not the case in dark field. On dark-field micrographs the asymmetry is the same whichever surface is involved. By taking micrographs of tracks in both bright and dark field one may therefore determine the point of entry of the particle into the crystal, and the direction in which it travelled. This is important information since the nature of energy loss changes as a charged particle *slows down*, so that variations in track structure (i.e. damage) should reveal something about the mechanisms of energy loss. *The direction* of the asymmetry of a dark-field image is also very important since it depends only upon the sign of B and the direction of **g**. Consequently it is only necessary to take micrographs of dark-field pictures of tracks in order to determine the sign of B for a given direction of $\mathbf{g}$. Hence, providing that the proper precautions are taken (Ashby and Brown 1963a, b), tracks which correspond to a cylindrical contraction (an inward curvature of the lattice) will show, on positive prints of dark-field micrographs, anomalously wide images with their 'white' side on the side of positive $\mathbf{g}$ and their 'black' side on the side of negative $\mathbf{g}$ (see Figure 38(c)). The reverse will be true for tracks which correspond to a cylindrical centre of compression. The important point here is that in determining the sign of B we are, in effect, saying something about the contents of the central cylinder. Thus, in the case of a contraction, the central cylinder might be considered to contain an excess of 'vacancies' while, for compression, 'interstitials' are prevalent.

The same kind of reasoning may be applied with success to circumstances where the tracks are not continuous but are intermittent. An example of this kind of track, observed in bright field, was shown in Figure 29. In this case the intermittent nature of the track was proved to be real by observing that the discontinuities were preserved on tilting the crystal in the stereo-device of the Siemens instrument. Now one possibility is that some of the intermittent tracks of the kind shown in Figure 29 could be due to the creation of damage by displacement cascades in a regime of energy loss where the damage produced by electronic excitation, and by the resulting thermal spike, is not very great. Brinkman, for example (Chapter 2), has pointed out that energetic primary knock-ons will be produced with increasing frequency as the fragment slows down (Brinkman, 1962). Each of these primaries will proceed to displace further atoms with a frequency which is related to the mean-free-path for displacement events

$$\lambda_d = 1/(n_0 \sigma_d) \tag{6.19}$$

where n_0 is the spatial density of lattice atoms, and σ_d is the total cross-section for displacement. When λ_d falls below the interatomic distance a_0, however, the displacements may no longer be considered separately since a multiple vacancy surrounded by many interstitials is formed. The resulting configuration, called a displacement, spike was illustrated schematically in two dimensions in Figure 8. The final result of the initiation of a displacement spike in a crystal lattice cannot be certain but, owing to the stirring effect of the displaced atoms as they are driven out of the central region and the effect of pressure and temperature on the volume bounded by the interstitial shell, it is unlikely that many of the atoms will return to their original sites. In a crystal containing one atomic species, Brinkman infers the existence of a number of Frenkel defects, regions of 'mismatch', and possibly dislocation loops. In a binary material, such as molybdenum disulphide, lead iodide, or nickel chloride, replacement events will dominate in producing disorder, and we are not able to say whether collapsed spikes (vacancy type) or expanded spikes (interstitial type) are likely to be formed (Bollmann, 1962).

Seeger and Essmann (1962) added an extra feature to the displacement argument. In a detailed discussion of the mechanism of radiation hardening in face-centred cubic metals they suggest that the central region of a displacement spike will be diluted of matter by the transport of energy and mass in the form of focusing collisions (Thompson, 1959; Thompson and Nelson, 1961) and dynamic conditions (Silsbee, 1957). Channelling (Holmes, 1963) may also be involved. The result is the so-called depleted zone (Figure 9), which, in contrast to the displacement spike, definitely does contain an excess of vacancies (see Chapter 2).

The anomalous nature of dark-field images of intermittent tracks may therefore be a powerful tool, since if the segments have a cascade origin the direction of the strain which may be deduced will permit an estimation of the validity of the depleted zone concept for the crystal in question. The mere fact that each segment of a track is of more complex symmetry than that of a cylinder or sphere (it is probably lenticular) is immaterial – the basic argument still applies.

We find in fact that in molybdenum disulphide the asymmetry of dark-field images (Figure 44) varies. Sometimes the strain corresponds to an excess of 'vacancies', and sometimes to an excess of 'interstitials'. We therefore conclude that, in a binary material at least, if the segments do indeed result from cascades initiated by the lower energy fragments, it is not necessary to add the extra feature of removal of matter far from the spike.[†] Replacement events are much more important in producing disorder and strain; collapsed spikes are discovered about as often as expanded spikes. It should be made clear, however, that in the event of intermittent tracks of atomic cascade origin being discovered in metals or elemental semiconductors it might be possible to adduce evidence for the concept of the depleted zone.

References

ASHBY, M. F., and BROWN, L. M. (1963a), *Phil. Mag.* **8**, 1083.

ASHBY, M. F., and BROWN, L. M. (1963b), *Phil. Mag.* **8**, 1649.

BETHE, H. A. (1928), *Ann. Phys. Lpz.* **87**, 55.

BOLLMANN, W. (1962), *Proc. Fifth Int. Congr. Electr. Microsc.*, Philadelphia, F-7, Academic Press (New York).

BOWDEN, F. P., and CHADDERTON, L. T. (1962), *Proc. Roy. Soc. A* **269**, 143.

BOWDEN, F. P., and CASPAR, P. E. (1963), *Phil. Mag.* **8**, 2091.

BRINKMAN, J. A. (1962), N. Amer. Aviation Rep. SR-6642.

CHADDERTON, L. T. (1962), *Nature*, Lond. **195**, 987.

CHADDERTON, L. T. (1963), *Phil. Mag.* **8**, 2085.

CHADDERTON, L. T. (1964), *Proc. Roy. Soc. A* **280**, 110.

HASHIMOTO, H., HOWIE, A., and WHELAN, M. J. (1962), *Proc. Roy. Soc. A* **269**, 80.

† A further and more realistic consideration relates to the possibility that intermittent track formation might be due to successive bursts of electronic energy loss from the higher energy fragments. This is discussed in Chapter 13 (page 213).

HIRSCH, P. B., HOWIE, A., and WHELAN, M. J (1960), *Phil. Trans. A* **252**, 499.

HOLMES, D. K. (1963), Summer School on Solid State Physics, Mol, Belgium.

HOWIE, A., and WHELAN, M. J. (1961), *Proc. Roy. Soc. A* **263**, 217.

HOWIE, A., and WHELAN, M. J. (1962), *Proc. Roy. Soc. A* **267**, 206.

LAME, G. (1852), *Leçons sur la Théorie de l'Élasticité*, Paris.

LOVE, A. H. (1927), *Mathematical Theory of Elasticity*, Cambridge University Press.

MOTT, N. F., and NABARRO, F. R. N. (1940), *Proc. Phys. Soc. Lond.* **52**, 86.

SEEGER, A., and ESSMANN, U. (1962), Enrico Fermi Summer School on Radiation Damage, Ispra, Italy.

SILK, E. C. H., and BARNES, R. C. (1959), *Phil. Mag.* **4**, 970.

SILSBEE, R. H. (1957), *J. Appl. Phys.* **28**, 1246.

THOMPSON, M. W. (1959), *Phil. Mag.* **4**, 139.

THOMPSON, M. W., and NELSON, R. S. (1961), *Proc. Roy. Soc. A* **259**, 458.

CHAPTER SEVEN

The Nature of Damage in Metals

So far we have dealt only with the *general* problem of fission damage in crystals, and the reason for presenting the material in this particular fashion has been twofold. Firstly, we wished to indicate the usefulness of methods which try to 'look at' the irradiated lattice 'directly'. Secondly, we wished to place the observations which have been made using one of these methods – electron microscopy – on a firm footing. We now turn to a more detailed discussion of the introduction of damage into *particular* kinds of material, beginning with the metals. Furthermore, since all of the metals (and also the elemental semiconductors, silicon and germanium) behave in a broadly similar way under similar conditions of fission fragment irradiation,† it will only be necessary to consider in detail the case of gold (Van Vliet and Chadderton, 1967). Discussion of the results in terms of damaging processes will be deferred to Chapter 14.

Specimen Preparation

Single-crystal films of gold suitable for electron microscopy were prepared by epitaxial growth following the methods of vacuum deposition described by Pashley (1959). Films grown by this method were preferred to foils electro-thinned from the bulk because the thickness is more accurately known and is easier to control.

About 2000 Å of silver were deposited onto a substrate of single-crystal rock-salt (NaCl), freshly cleaved along a (100) face, and heated above 250°C. This was followed immediately by the required thickness of gold. The silver grows epitaxially as a continuous single crystal which then acts as the substrate for the subsequent deposition of gold. Gold films grown directly onto rock-salt tend to be polycrystalline and noncontinuous but, by growing them on the intermediate layer of silver, continuous single-crystal films down to 100 Å thick can be obtained. Both the gold and the silver grow with a $\langle 001 \rangle$ film normal.

† This has been confirmed repeatedly by electron microscopy of several metals and by field-ion microscopy (Appendix 1) of tungsten. Alloys, in spite of their compound nature, behave likewise (Van Vliet, 1966).

Single crystals of rock-salt, 1 cm × 1 cm cross-section, were carefully selected, and the substrates were sections less than 1 mm thick cut by cleaving with a razor blade. These were heated under vacuum on an insulated tantalum disc in the centre of a vertical oven. The oven consisted of a 1 in. diameter cylinder of 0·001 in. tantalum foil and was resistance heated. The temperature of the substrate was measured by a thermocouple placed near it, and was controlled manually. A hot-plate heater of the type suggested by Pashley was tried initially, but was abandoned since it was found to be difficult to obtain a

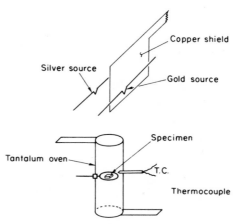

Figure 45. A sketch of the substrate oven and the evaporation geometry for the preparation of gold film.

good thermal contact between the substrate and the plate. The oven method had the disadvantage that outgassing from the oven created a local increase in pressure near the substrate, but this did not appear to affect the results. All evaporations were at a nominal pressure of better than 1×10^{-5} torr.

The gold and silver were evaporated downwards from tungsten hairpin filaments held about 8 cm from the substrate and separated by a copper shield. The best films were obtained with a substrate temperature of 280°C, and evaporation rates for both silver and gold of between 200 and 500 Å/min. Figure 45 shows a sketch of the oven and evaporation geometry.

The gold film was detached by successively dissolving away the rock-salt and silver in baths of distilled water and nitric acid. The gold film was then cleaned in distilled water and picked up on electron microscope grids. Up to twenty specimens could be obtained from a single 1 cm × 1 cm film.

The film thickness was calculated by fully evaporating a measured amount of metal, but due to non-isotropic evaporation from the filament it was only accurate to within 20%. Roughly the same accuracy could be achieved, however, by measuring the widths of stacking faults and slip traces in the films, and the two methods were found to give good agreement. In all actual experiments the thickness was determined by the first method.

Gold films obtained in this manner often contained a high density of dislocations (up to $10^{10}/cm^2$), stacking faults, and microtwins. On flash-annealing the specimens in a tantalum boat under vacuum at 700°C for up to 30 sec, most of the dislocations annealed out and the microtwins grew longer – sometimes up to several microns. This made the post-irradiation analysis simpler as it removed most of the 'background' defects. Sometimes on flash-annealing, the films appeared to melt and recrystallize with a (111) orientation, and quite often 'half-and-half' films were obtained. It appeared that no change in thickness accompanied either the normal flash-annealing or recrystallization.

Electron Microscopy

After irradiation with random fission fragments in the manner described previously (Chapter 3) specimens were examined in an AEI EM 6 electron microscope operating at either 75 or 100 keV. Most micrographs were taken at a nominal magnification of 100,000. The actual magnification was checked periodically by measuring the diameters of circular latex particles of known size (usually $0·557 \pm 0·011$ diameter as supplied by Dow Chemical Company). To minimize variations in magnification all micrographs were taken under as nearly identical microscope conditions as possible. For example the final magnification was always reached by increasing the excitation current of the first projector lens *up* to its final value, so that the effect of hysteresis in the lens core was always the same.

In order to study the character of defects using diffraction contrast theory it is essential to operate the electron microscope under 'dark-field' conditions, i.e. with the image formed by a diffracted beam instead of the transmitted beam. A detailed survey of diffraction contrast theory and dark-field techniques is given in *The Electron Microscopy of Thin Crystals* (Hirsch, Howie, Nicholson, Pashley, and Whelan, 1965). We sketch a few of the details which apply specifically to the AEI instrument.

To obtain a dark-field image of high resolution the image-forming diffracted beam must pass as nearly as possible down the optic axis of the objective lens so as to minimize spherical aberration. Now to bring a diffracted beam into coincidence with the optic axis it must be tilted about an axis normal to itself. This axis is the line of intersection of the horizontal plane of the microscope with the crystal planes in the specimen which diffract the beam. The appropriate axis of tilt is thus determined by the specific orientation of the specimen, and it follows that any rotation of the specimen in the horizontal plane also rotates this axis by the same amount.

Since any tilt applied to the incident electron beam is also applied equally to all diffracted beams, tilting the incident beam about the appropriate axis also tilts the diffracted beam in the required direction. In the EM 6 the incident beam may be tilted electromagnetically between the second condenser lens and the specimen but only about one fixed horizontal axis. It is obviously necessary to bring the desired axis of tilt into coincidence with the fixed axis. This may be accomplished by rotating the specimen in the horizontal plane using a special rotating holder supplied by AEI. A lower half containing the specimen in a cap can be rotated about the upper half. Since the position of the horizontal support rods is fixed in the specimen chamber this is equivalent to rotating the specimen relative to the microscope, and with the proper rotation the two axes mentioned above may be brought into coincidence.

To determine the required rotation the specimen is inserted in the microscope, and its diffraction pattern imaged on the screen. Tilting the beam displaces the pattern uniformly, and obviously any spot which lies on a straight line through the optic centre of the screen in the direction of this displacement can be moved to the centre. Under these conditions the beam producing the spot is passing down the optic axis of the objective lens. Since the diffraction pattern rotates directly with the specimen, the rotation required to bring a diffraction spot onto this line may be estimated. The specimen holder is removed from the microscope, rotated accordingly, reinserted and the diffraction pattern checked again. Usually the proper rotation can be found after only two attempts. Once this has been achieved the beam is tilted through the necessary angle, and the microscope is ready for high-resolution dark-field work.

The size and density of the visible defect clusters which were observed in irradiated specimens were measured quickly and accurately by projecting the micrograph plates onto a sheet of white paper with a photographic enlarger. In determining relatively high cluster densities a correction had to be made to account for the overlap of images from clusters at different depths in the film. A simple analysis shows that the true density, N_T, is related to the observed density, N_c, by:

$$N_T/N_c = 1 + \tfrac{1}{2}N_c A + \tfrac{1}{3}(N_c A)^2 \tag{7.1}$$

where A is the mean image size. The ratio N_T/N_c was always less than 1·2.

In determining image diameters all measurements were made on micrographs taken in bright field under kinematic conditions (i.e. at areas far from any extinction contours) so that the effect of variable diffraction contrast conditions was minimized. In the case of non-circular images the *shortest* diameter was taken.

General Observations

The effects of fission fragment (FF) irradiation on gold as observed by bright-field microscopy may be summarized as follows:

(1) At low doses (less than 5×10^{11} FF/cm² for 3 h using natural uranium) damage consisted of small black spots between 20 and 100 Å in diameter, qualitatively similar to those commonly observed in other irradiated metals.

The density† of visible clusters over this range of doses was directly proportional to the total dose. As the dose was increased the proportion of large clusters increased, and the rate of increase of the density of clusters with dose became slower – a logarithmic rather than a linear relationship was discovered. At very high doses (above 10^{13} FF/cm^2 for 30 min using enriched uranium) short lengths of dislocation lines formed. Although films were irradiated with up to $1 \cdot 3 \times 10^{14}$ FF/cm^2 ($6\frac{1}{4}$ h using enriched uranium) no long-range dislocation networks were observed. The qualitative variation of the damage with dose is illustrated in Figure 46(a) to (d). A more quantitative discussion of the variation of the size and density of the clusters with dose is given below.

(2) Continuous tracks or linear arrangements of clusters were not observed even for high doses. However, groups of anomalously closely spaced spots were occasionally found even at very low doses (Figure 47).

(3) Specimens given an equivalent total dose from natural or enriched uranium showed no variation within experimental error of either the size or concentration of clusters. Increasing the flux by a factor of 135 did not affect the results.

(4) Denuded areas near grain boundaries, microtwins, stacking faults, dislocations, or holes in the film were not observed. A comparison of the density of clusters in 150 Å and thicker films indicated that, if denuded layers exist near the surfaces, they are almost certainly less than 20 Å thick.

(5) The densities of clusters in a 700 Å and a 350 Å thick film were measured after a total dose of $3 \cdot 2 \times 10^{11}$ FF/cm^2 (2 h using natural uranium), and their ratio was found to be approximately 2:1. For films much thicker than any denuded surface layers, it appears that the density of clusters is directly proportional to the thickness, or that the concentration is independent of thickness.

(6) Sometimes in (001) films rectilinear images were observed – generally squares with sides parallel to $\langle 100 \rangle$ directions in the plane of the film (Figure 48). Later we shall identify these images as those of stacking fault tetrahedra similar to those which are commonly observed in quenched and annealed gold foils (Silcox and Hirsch, 1959). We shall also show that clusters in pre-irradiated specimens may be transformed into tetrahedra upon annealing. Since tetrahedra are not found in all irradiated specimens, we feel that their formation may be caused by abnormally high irradiation temperatures which are known to occur during reactor operations.

(7) During observation in the microscope, clusters in (001) films were occasionally observed to oscillate rapidly between two positions separated by distances of the order of 1000 Å. In all cases the direction of motion was parallel to a planar $\langle 110 \rangle$ direction. Since the films were only a few hundred Ångströms

---

† It should be noted that 'density' refers to the number of clusters per unit surface area, while 'concentration' refers to the number per unit volume. Because of the possibility of a denuded surface layer the ratio of the two is not necessarily the film thickness. To avoid this problem we generally used specimens of identical thickness, so that the concentration is always proportional to the directly measured density.

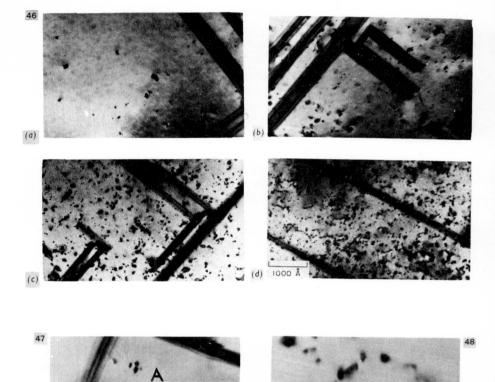

Figure 46. A sequence of four micrographs illustrating the effects of increasing the total fission fragment dose. All specimens have been taken from the same 350 Å thick, (001), flash annealed, gold film specimen 'B'. (a) Irradiated with 4.0×10^{10} FF/cm² (15 min using natural uranium), a relatively low dose. Almost all clusters are characterized by small black spot images. (b) 2.0×10^{11} FF/cm² (75 min using natural uranium). The proportion of large clusters has increased very slightly. (c) 1.0×10^{12} FF/cm² ($6\frac{1}{4}$ h using natural uranium). The clusters have begun to overlap in several cases. (d) 2.7×10^{13} FF/cm² (75 min using enriched uranium). Note the formation of short lengths of dislocation lines.

Figure 47. Electron micrograph of another area of the same (001) gold specimen shown in Figure 46 (specimen B) after a fission fragment dose of only 8×10^{9} FF/cm² (3 min using natural uranium). Note the groups of several closely spaced clusters at A, and the very low mean cluster concentration elsewhere.

Figure 48. A 300 Å, (001), gold film irradiated to a total dose of 4×10^{12} FF/cm² (10 min using enriched uranium). The microtwin shown on the right-hand side is parallel to a $\langle 110 \rangle$ direction. Note the square images at A with sides parallel to planar $\langle 100 \rangle$ directions associated with stacking fault tetrahedra.

thick this motion must be along a crystallographic $\langle 110 \rangle$ direction in the plane of the film, and not along an oblique direction which projects onto $\langle 110 \rangle$. The number of clusters showing this oscillatory motion was less than 10% of the total. Such oscillating clusters are almost certainly perfect prismatic dislocation loops which glide along cylinders parallel to their Burger's vector $(\vec{b} = \frac{1}{2} \langle 110 \rangle)$ and lie in the plane of the film. A similar glide of faulted loops is not possible. The same observations and interpretation have been made by Ruedl *et al.* (1962) on fission fragment irradiated platinum, and by Pashley and Presland (1960) on irradiated gold.

(8) The damage caused by the neutron flux, which contained fast as well as thermal neutrons, was negligible. A 350 Å thick gold film irradiated without a uranium backing for 10 h in BEPO, a nominal dose of about 1×10^{16} fast neutrons/cm^2, had a cluster density less than that observed in a similar specimen irradiated for only 3 min with natural uranium backing, a total dose of 8×10^9 FF/cm^2.

The Variation of Cluster Density with Dose

To obtain more quantitative information on the rate of formation of clusters with dose, two specimens, A and B, were irradiated for 3, 15, 75, and 375 min with both natural and enriched uranium. Both were nominally 350 Å thick. A was irradiated in the as-grown condition, while B had been flash-annealed. In Figure 49 the observed cluster density is plotted against the total fission fragment dose on a logarithmic/logarithmic scale. The dotted reference line represents a slope of unity. For both specimens A and B a linear relationship holds up to a cluster density of approximately 5×10^{10} clusters/cm^2. In this region the number of visible clusters per fission fragment striking the film is 0·6 for the as-grown specimen and 0·2 for the flash-annealed specimen. The reason for this large discrepancy is not clear, though it may well be due to larger impurity concentrations in the annealed specimens which retard the formation of clusters. The fact that the cluster density approaches the same values at high doses indicates that the effect is probably not due to any radical change in thickness upon flash-annealing.

From Figure 49 the cluster density appears essentially to saturate at doses above 1×10^{12} FF/cm^2 (3 min using enriched uranium). If the density is now plotted against the logarithm of the total dose in this region (see Figure 50) a good linear fit is obtained. This result should be treated with some caution, however, since experimental errors are of the same order as the differences involved.

The Variation of Image Size with Dose

The mean image diameter was measured for (111) and (001) oriented films (160, 350, and 470 Å thick) for irradiations with natural uranium of 3, 15, 75,

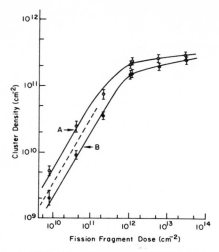

Figure 49. The density of visible clusters plotted against the total fission fragment dose on a logarithmic/logarithmic scale for different irradiated gold specimens. Both were nominally 350 Å thick – A was as-grown, and B had been flash annealed. The concentration of clusters is directly proportional to the density in both cases. The dotted line has a slope of unity and represents a linear relationship.

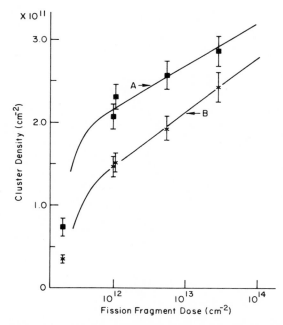

Figure 50. The density of visible clusters plotted against the logarithm of the total fission fragment dose for specimens A and B (see Figure 49) at relatively high doses.

and 375 min. The results are given in Tables II and III, where values for the 160 and 350 Å films are the average of two different specimens. The mean image diameters appear to be independent of thickness but to increase very slowly with dose. Although the quantity which has been directly measured is the image diameter, not the diameter of the cluster itself, the same conclusions apply equally well to the mean cluster diameter.

The fact that the mean image diameters are greater in (111) films than in (001)

TABLE II
Mean image diameter in (001) gold films (Å)

Thickness (Å)	Total dose (FF/cm^2)			
	$8 \cdot 1 \times 10^9$	$4 \cdot 0 \times 10^{10}$	$2 \cdot 0 \times 10^{11}$	$1 \cdot 0 \times 10^{12}$
160	50	47	50	50
350	45	48	53	50
470			53	54

TABLE III
Mean image diameter in (111) gold films (Å)

Thickness (Å)	Total dose (FF/cm^2)			
	$8 \cdot 1 \times 10^9$	$4 \cdot 0 \times 10^{10}$	$2 \cdot 0 \times 10^{11}$	$1 \cdot 0 \times 10^{12}$
160	39	53	55	58
350	56	54	59	61
470	49	55	58	59

films is probably due to differences in the diffraction conditions rather than an orientation dependence of the cluster size. As a general rule the image of a small loop or spherical strain-field increases in size as the extinction distance of the operating diffraction vector increases. Since the shortest possible extinction distance in a (001) film is $\xi_{200} = 170$ Å (at 100 keV), while in a (111) film it is $\xi_{220} = 236$ Å, the images should on the average be somewhat larger in (111) films.

The distribution of image diameters was found for several different films, and typical normalized histograms are shown in Figure 51. The approximated curves are qualitatively similar in both examples, although the specimen in Figure 51(a) received a dose 5 times greater than that in Figure 51(b).

Although no images of diameter less than 20 Å were measured, this is thought to be due to a visibility limit in the microscope rather than an inherent minimum associated with the clusters. Hesketh and Rickards (1966) in work on thin copper foils bombarded with 150 keV Zn^+ ions find a minimum image size which they associate with clusters of about 270 vacancies. Since these investigators determined the image sizes from microdensitometer traces and employed more exact diffraction conditions, their results are no doubt more accurate than those described here. We nevertheless feel that our results would indeed show such a large minimum cluster size if it did in fact exist.

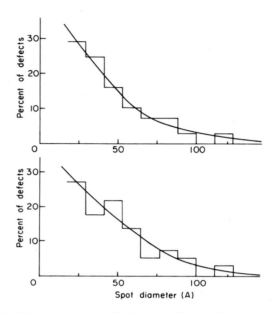

Figure 51. Normalized histograms of the distribution of image diameters for two different total doses. (a) A 470 Å thick, (001), gold film irradiated to a dose of $1 \cdot 0 \times 10^{12}$ FF/cm² (6¼ h using natural uranium) 69 images measured. (b) A 160 Å thick, (001), gold film irradiated to a dose of $2 \cdot 0 \times 10^{11}$ FF/cm² (75 min using natural uranium) 96 images measured.

Analysis of the Clusters

It should be possible to use the two-beam electron diffraction contrast theory to establish the crystallographic structure of defect clusters from an interpretation of the characteristics of their images as observed in the electron microscope. Thus we may be able to determine whether they are vacancy or interstitial type defects. Are they spherical clusters of closely spaced point defects, voids, or dislocation loops? If loops, what are their Burger's vectors and habit planes, and are they rectilinear or circular in shape (i.e. bounded by dislocations along low index crystallographic directions or by a heavily jogged dislocation line)?

If the clusters are indeed loops, they are clearly so small that it is impossible to distinguish dislocation lines bounding the loop or to observe whether the loop is faulted or not. The image of such small defects is determined by the strain field set up by the dislocation loop as a whole – small segments of the bounding dislocation lines cannot be considered separately. To interpret the images from small loops or clusters we must use the analyses of the two-beam electron diffraction contrast theory.

Ashby and Brown (1963a) have calculated the image of an *inclusion* producing a spherically symmetrical radial strain field. Inclusions within about one extinction distance of either foil surface have an asymmetric elliptical image with characteristic 'black' and 'white' sides along the major axes. As in the case of fission particle tracks (Chapter 6) the direction of black/white streaking is always parallel to $\pm\mathbf{g}$ (the reflection vector in reciprocal space) under two-beam conditions, the sign depending on the character of the strain field (vacancy or interstitial type) and on whether the inclusion is near the top or bottom surface of the foil. In dark field the 'sense' of the asymmetric image is the same for inclusions near either surface, whereas it reverses in bright-field microscopy. An asymmetric dark-field image always has its white side on the side of positive $\mathbf{g}$ if the inclusion is vacancy type, and on the side of negative $\mathbf{g}$ if it is interstitial type. Dark-field images can thus distinguish uniquely between the two cases. This rule, however, is found to apply only to relatively large inclusions. For small inclusions (e.g. cobalt precipitates in copper less than 200 Å in diameter (McIntyre and Brown, 1966)) the sense of asymmetric images may indicate either a vacancy or interstitial type strain field. The rule that the direction of streaking is always parallel to $\pm\mathbf{g}$ appears to be perfectly general and applicable to all defects possessing a spherically symmetrical strain field.

The calculations have been extended to the images of small dislocation loops by Ashby and Brown (1963b), McIntyre and Brown (1966), and Ruhle, Wilkens, and Essman (1965). The situation now is much more complicated. The black/white image is still expected for loops near a foil surface, but the direction of streaking depends on the operating reflection vector, the Burger's vector, and the normal to the plane of the loop. Moreover the sense of the black/white image varies with the distance of the loop from a foil surface. Using the 'infinitesimal loop approximation' the direction of maximum black/white streaking appears to be principally determined by the direction of the Burger's vector **b** (or by the projection of **b** onto the image plate) with a possible shift of up to 10° toward **g**. Although some workers associate with the observed black/white images a line of no contrast not necessarily normal to the direction of maximum streaking or **g**, the detailed calculations are unable to deal with such an effect since the infinitesimal loop approximation breaks down for imaging very near the centre of the loop. Conclusions drawn from such observations must be treated with some reserve.

Dark-Field Microscopy

Cluster images were studied in dark field using (200) and (220) reflections from (001) films and ($2\bar{2}0$) reflections from (111) films. Because of the described difficulties involved in aligning the EM 6 electron microscope for high-resolution dark-field microscopy the same clusters were not imaged with different diffracted beams. Almost all images observed in dark-field microscopy showed the expected black/white contrast with well-defined streaking.

We discuss first the results of (001) films. Viewed in (220) dark field the direction of black/white streaking was almost always parallel to $\pm[110]$, i.e. $\pm\mathbf{g}$. Approximately 5% of the clusters had black/white images along [100] or [010]. When the same specimen was imaged using a (200) diffracted beam, although many images still streaked along [110] or [1$\bar{1}$0], about one-third of the black/white images were now approximately parallel to $\pm[100]$, i.e. $\pm\mathbf{g}$. This tendency was most pronounced for the smallest images. Moreover the observed density of clusters was a factor of approximately 2·0 greater using a (200) diffracted beam than a (220) beam.

What interpretation may be given to these results? Firstly, it would seem that those images which have black/white streaking along either [100] or [010] when observed with a (220) reflection can only be due to perfect dislocation loops with a Burger's vector ($\frac{1}{2}\langle101\rangle$) obliquely inclined to the plane of the film. All other possible Burger's vectors for dislocation loops ($\frac{3}{4}\langle111\rangle$ or $\frac{1}{2}\langle110\rangle$) have projections which lie along either [110] or [1$\bar{1}$0]. In the latter case they are normal to g and are not visible. Moreover, any cluster possessing a spherically symmetrical strain field should have an image with black/white streaking parallel to $\pm\mathbf{g}$ (i.e. $\pm[110]$). Since, as we shall indicate below, the actual number of visible clusters is perhaps twice the number observed here, the percentage of perfect loops with **b** inclined to the film is certainly less than 5%. We distinguish between perfect loops with **b** in the plane of the film, and those with an oblique **b**, since the latter may easily disappear by gliding out to a surface.

If we accept that very few loops have Burger's vectors projecting onto $\langle100\rangle$ directions in (001) films then those images with black/white streaking parallel to [100] for $\mathbf{g} = (200)$ would appear to arise from defects with a spherically symmetrical strain field. The degree of alignment with g is, however, far less perfect than for (220) dark-field microscopy, and it is quite conceivable that the effect is entirely due to that shift of the streaking direction away from **b** towards g mentioned above. At the most, therefore, the fraction of defects with a spherical strain field must be less than one-third. Those clusters whose images have black/white streaking along [110] or [1$\bar{1}$0] when $\mathbf{g} = (200)$ must almost certainly be dislocation loops – either perfect loops of planar Burger's vector or Frank loops whose Burger's vector ($\mathbf{b} = \frac{1}{3}\langle111\rangle$) projects onto a planar $\langle110\rangle$ direction.

The variation in cluster density observed upon changing the operating reflection vector **g** also indicates that most of the clusters must be loops. The fraction of all loops in (001) films that should be visible using (200) and (220) diffracted beams has been calculated for both Frank and perfect loops. We have assumed that all equivalent Burger's vectors are equally probable, and that a loop is visible only if the angle between the projection of **b** and $\pm$**g** is 45° or less (experimentally this appears to be justified). We again distinguish between perfect loops of planar and oblique **b** for the reason given above. The results are given as ratios in Table IV. Approximately equal fractions of all spherical clusters in (001) films are visible when imaged in either (200) or (220) dark-field microscopy. The minimum resolvable loop or cluster size is roughly proportional to $(\xi_g/g)^{1/2}$ (see Figure 15, Ashby and Brown, 1963a) which is not too different for **g** = (200) or (220). The image sizes may change but the visibility criterion does not. This result is also listed as a ratio in Table IV. Experimentally the ratio of the cluster density observed using a (200) reflection to that observed with a (220) reflection was approximately 1:0·5. From Table IV it appears that this ratio is most consistent with clusters consisting of either Frank loops or perfect loops of planar Burger's vector, in agreement with our conclusions above.

TABLE IV

Fraction of clusters visible in (001) gold films for different **g**

Diffraction vector	Frank loops	Perfect loops		Spherical clusters	Experimental ratio
		planar **b**	oblique **b**		
(220)	0·5	0·5	1·0	1·0	0·5 ± 0·1
(200)	1·0	1·0	0·5	1·0	1·0

Dark-field microscopy of (111) films can partially distinguish between Frank and perfect loops since the projection of a $\langle 111 \rangle$ direction onto the plane of the film is now 30° away from a planar $\langle 110 \rangle$ direction. The smaller images tend to have a black/white streaking parallel to $\pm$**g**. This was also noted for (001) films with **g** = [200], and the same explanation, if valid, should be equally applicable here. A statistical count on only the relatively large images shows that roughly three times as many images have a black/white streaking direction which lies $\pm 30°$ with $\pm[1\bar{1}0]$ as those which streak along $[1\bar{1}0]$. In other words three times as many of these images are typical of Frank loops than are typical of perfect loops of planar Burger's vector. Taking account of the fraction of Frank and planar perfect loops that are visible under these conditions (0·5 and 0·33 respectively), the perfect loops of planar Burger's vector constitute less than 30% of the total concentration of clusters. Owing to the tendency of black/white streaking to shift toward **g** this is probably an overestimate and 20% may be a more reasonable figure. This would be in rough agreement with the percentage of clusters (<10%) which were observed to oscillate in the microscope beam and which were associated with perfect loops of planar Burger's vector.

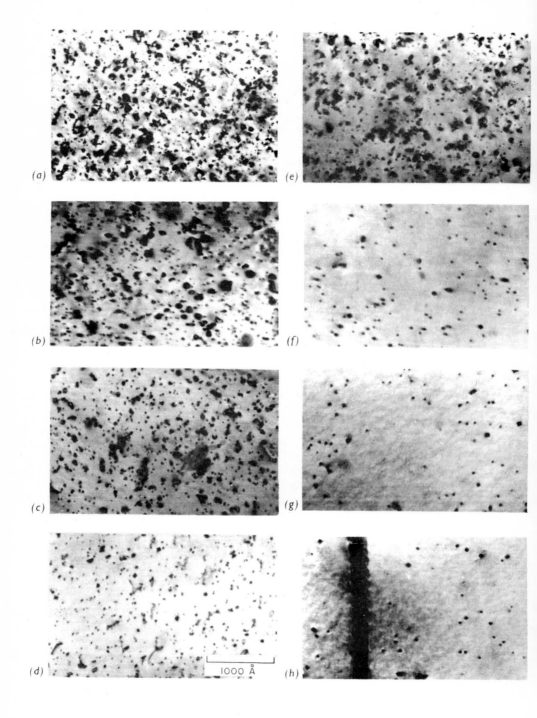

(a)

(b)

(c)

(d)

(e)

(f)

(g)

(h)

1000 Å

Having established that most of the clusters are Frank loops an estimate may be made of the number of point defects per loop. We assume that the measured image diameters are approximately equal to the loop diameters (in fact, of course, they are overestimates), and that the loops are circular. On this basis, clusters whose image diameters range from 20 to 100 Å contain from 40 to 1100 point defects.

We note in passing that calculations by Sigler and Kuhlmann-Wilsdorf (1966) show that the lowest energy configuration of a vacancy cluster in gold is a void for less than ten vacancies, and a perfect loop for over 50,000. For clusters of an intermediate size a Frank loop or a stacking fault tetrahedron has the lowest energy, but an appreciable activation energy is involved in going from a Frank loop to a tetrahedron. Hence the observed preference for Frank loops is also expected on a theoretical basis.

Whenever statistical counts were made to determine whether the black or white side lay more frequently in the general direction of positive $\mathbf{g}$, the results were always, within experimental error, 50%. This result was independent of film orientation and operating vector, and the same whether all defects or only the largest were counted. It therefore appears to be impossible in this case to distinguish between vacancy and interstitial type clusters by the sense of the black/white asymmetry without the use of more sophisticated techniques.

The Annealing of Clusters

In this section we describe some experiments on post-irradiation annealing of the clusters in fission fragment irradiated gold. These were performed in order that, by a comparison with the annealing behaviour of known defects, the character of the clusters might be established. In particular it was hoped that it might prove possible to distinguish between the possibilities that the clusters might be all vacancy, all interstitial, or a combination of both. The irradiated specimens were annealed under a vacuum of better than 1×10^{-4} torr in a quartz tube inserted in a wire-wound controlled furnace. The annealing

---

Figure 52. A sequence of four micrographs illustrating the effect of isothermal annealing at 250° on a 350 Å thick, (001), gold film irradiated to a dose of $3 \cdot 6 \times 10^{12}$ FF/cm² (10 min using enriched uranium). Note the gradual disappearance of the larger loop-type images. The total concentration of clusters decreases only slightly throughout the sequence. (a) The as-irradiated specimen. (b) Annealed for 30 min at 250°C. (c) The same specimen annealed for 90 min at 250°C. Note the formation of tetrahedra. (d) Annealed for 335 min at 250°C. Also a sequence of four micrographs showing the effect of isothermal anneals at 350°C. The specimen was 350 Å thick, (001) orientated, and had been irradiated to a dose of $1 \cdot 1 \times 10^{12}$ FF/cm² (3 min using enriched uranium). (e) The as-irradiated specimen. (f) Annealed for 45 min at 350°C. Although some 'black spots' still remain, almost all the clusters now have the square images characteristic of tetrahedra. The concentration of clusters has been reduced to approximately one-third the original, and the lower annealing stage is virtually complete. (g) The same specimen annealed for 165 min at 350°C. All the remaining clusters are now tetrahedra. (h) Annealed for 225 min at 350°C. Very little change is to be noted from (c). A (220) dark-field micrograph of this same annealed specimen is shown in Figure 24.

characteristics were divided into two stages – a lower one operative from about 250 to 350°C and an upper one above 350°C.

The Lower Annealing Stage (250 to 350°C)

The effect of isothermal annealing at 250°C may be seen in Figures 52(a) to (d). The major change was a gradual decrease in the size of the larger clusters such that, after about 5 h, almost all the remaining clusters had images characterized by black spots. The total concentration of clusters decreased only very gradually during annealing at 250°C, while many tetrahedra were observed to form.

The initial effect noted on isothermal anneals at 300°C was also a shrinking of the larger clusters. This was, however, accompanied by a definite decrease in the cluster concentration until, after 5 h, the concentration appeared to approach an equilibrium value. Tetrahedra were formed at a slightly higher rate than on anneals at 250°C.

At 350°C the equilibrium value was reached in about 10 min, and thereafter the annealing rate was much slower (Figures 52(e) to (h)). The latter annealing is thus classified as part of the upper stage. After a 1 h anneal at 350°C, annealing in the lower stage was judged to be complete. Most clusters remaining had images typical of tetrahedra, although a varying proportion still appeared as simple black spots.

A number of different specimens were annealed for 1 h at 350°C and examined in the electron microscope. In all cases the concentration of clusters which survived the lower annealing stage was between 30% and 50% of the original. The wide scatter was within experimental limits. This effect was independent of film thickness, total irradiation dose, flux, or prolonged pre-annealing at 250°C. It appeared to be associated with some inherent characteristic of the clusters themselves, rather than a surface or annealing rate dependent effect.

The Upper Annealing Stage (350 to 650°C)

The slow annealing rate already mentioned for the latter part of isothermal anneals at 350°C has been associated with the upper stage. It is apparent from Figures 52(g) and (h) that almost all remaining clusters have the characteristic square shape of tetrahedra and the upper stage appears to be uniquely associated with this defect.

30 min isochronal anneals at 50°C intervals from 450 to 650°C indicated that most tetrahedra anneal out below 500°C, but that some are still present at 600°C (Figures 53(a) to (d)). Although no statistical counts have been made, a visual examination of many micrographs indicates that the smaller tetrahedra

anneal out at lower temperatures. This is quite evident in Figures 53(c) and (d), where the surviving tetrahedra are of relatively large dimensions compared to Figure 53(b).

In summary we conclude that two stages may be associated with the annealing of irradiated gold: a lower one in which about 60% of the clusters anneal out, while the rest transform to tetrahedra, and an upper stage associated solely with tetrahedra.

The Identification of Stacking Fault Tetrahedra

So far we have implicitly assumed that the square defects observed in either the as-irradiated or the annealed (001) films are stacking fault tetrahedra. We now show that this interpretation is consistent with the observations.

TABLE V

Face	Indices	$\mathbf{R}$	$\alpha = 2\pi\,(220).\,R$
ABC	$(\bar{1}\bar{1}1)$	$\frac{1}{6}[112]$	$2\pi/3$
BCD	(111)	$\frac{1}{3}[111]$	$\pi/3$
ABD	$(1\bar{1}\bar{1})$	$\frac{1}{6}[211]$	0
ACD	$(\bar{1}1\bar{1})$	$\frac{1}{6}[121]$	0

A stacking fault tetrahedron consists of stacking faults on the four non-equivalent {111} planes arrayed in the shape of a tetrahedron or pyramid and bounded by six stair-rod dislocations along ⟨110⟩ directions. Figure 54 illustrates the projection of a tetrahedron onto the (001) plane. The outer sides are the projections of the four inclined ⟨110⟩ directions onto planar [100] or [010] directions. Thus tetrahedra viewed in (001) films have a basically square image with sides parallel to ⟨100⟩, in accordance with our observations.

Moreover, when viewed in dark field using a reflection vector $\mathbf{g} = (220)$ only two of the tetrahedral faces are in contrast. The reason for this is obvious from Table V which lists the stacking fault vectors $\mathbf{R}$ associated with each of the tetrahedral faces of ABCD in Figure 54 and the 'phase' angle $\alpha = 2\pi\mathbf{g}.\,R$. If α is an integral multiple of 2π there is no phase shift in the diffracted beam from one side of the fault to the other and hence no change in the intensities of the transmitted and diffracted beams. Thus if $\mathbf{g} = (220)$ only the faces ABC and BCD are in contrast. Since the edge lengths are less than the extinction distance ($\xi_{220} = 236\ \text{Å}$) no stacking fault fringes are expected and the tetrahedral faces are either uniformly lighter or darker than the background. The image should consist of a pair of 'in contrast' triangles with their common base normal to $\mathbf{g}$. By symmetry the common base must be a line of no contrast. In Figure 55, a (220) dark-field micrograph of an annealed specimen, many of the larger images show these expected features. Thus the square images observed in

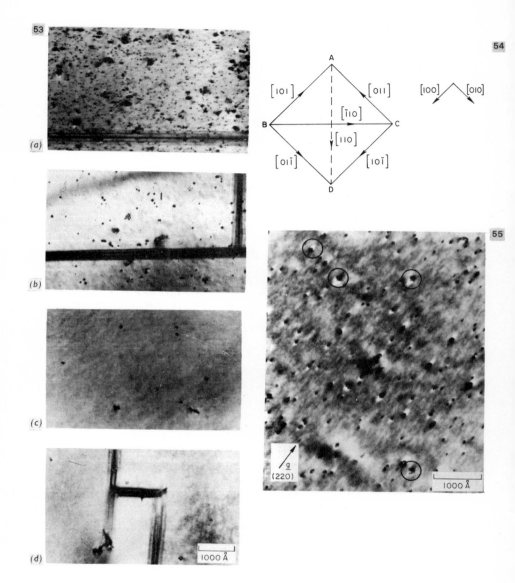

53

(a)

(b)

(c)

(d)

54

$[101]$ $[011]$

$[\bar{1}10]$

B $[110]$ C

$[01\bar{1}]$ $[10\bar{1}]$

$[100]$ $[010]$

55

$\underline{g}$
(220)

1000 Å

1000 Å

annealed (001) gold films must correspond to stacking fault tetrahedra, and similar images in as-irradiated films must almost certainly arise from tetrahedra.

The number of point defects contained in the tetrahedra may be calculated. This is equal to the number of defects contained in a Frank sessile loop of area equal to that of one tetrahedral face. The observed tetrahedra had edge lengths varying from 50 to 175 Å, and hence contain between 150 and 1700 point defects. The upper limit is somewhat larger than our original experimental estimate of a maximum of 1100 point defects per cluster. However, the former estimate was based on the sizes of clusters observed after a relatively low dose, while the tetrahedral edge lengths were measured in films given much higher doses. In these as-irradiated films an appreciable concentration of clusters with image diameters over 100 Å was present. The lower limit was difficult to determine since it was impossible to identify unambiguously any tetrahedra of edge lengths less than 50 Å.

It should be noted here that although interstitial stacking fault tetrahedra have never been positively identified (those in quenched metals must be vacancy type) there appears to be no inherent reason why they should not exist. There may however be important differences between the formation mechanisms of vacancy and interstitial stacking fault tetrahedra. Silcox and Hirsch (1959) originally proposed that tetrahedra grew from Frank sessile loops. When a Frank dislocation in the (111) plane lies along a $\langle 1\bar{1}0 \rangle$ direction it splits into a stair-rod dislocation and a Shockley partial in the intersecting $(11\bar{1})$ plane by a reaction of the type:

$$\tfrac{1}{3}[111] \rightarrow \tfrac{1}{6}[110] + \tfrac{1}{6}[112] \tag{7.2}$$

The angle between the plane of the loop and the plane of the partial must be acute for equation 7.2 to be energetically favourable. If the loop dissociates along three $\langle 1\bar{1}0 \rangle$ directions parallel to the sides of an equilateral triangle, this condition implies that all three partials lie on the same side of the loop plane

---

Figure 53. A sequence of micrographs illustrating the effect of isochronal anneals on tetrahedra. All four are of specimens taken from the same (001), 350 Å thick, gold film irradiated to a dose of 3·2 × 10¹¹ FF/cm² (2 h using natural uranium). (a) An as-irradiated specimen. (b) Annealed for 75 min at 350°C. The lower annealing stage is now complete; the clusters remaining are all tetrahedra. (c) An irradiated specimen annealed 30 min at 500°C. Note that the concentration of tetrahedra hàs decreased markedly from (b), and that those remaining are relatively large. (d) An irradiated specimen annealed 30 min at 600°C. Virtually no tetrahedra are left; only the very largest ones still remain.

Figure 54. The projection of a tetrahedron ABCD onto the (001) plane.

Figure 55. A (220) dark-field micrograph of the same irradiated and annealed specimen shown in Figure 21(d), which has been irradiated to a dose of 1·1 × 10²¹ FF/cm², and annealed for 225 min at 350°C. The circled images show the 'pair of in-contrast triangles' expected from stacking fault tetrahedra. The black/white streaking is due to the strain in the lattice surrounding the tetrahedra.

and are inclined inward. The partials attract each other by their elastic interaction and glide forward leaving a plane of stacking faults behind them. Along their line of intersection two Shockley partials combine to give a stair-rod dislocation by a reaction of the type:

$$\tfrac{1}{6}[112] + \tfrac{1}{6}[\overline{1}2\overline{1}] \rightarrow \tfrac{1}{6}[0\overline{1}1] \qquad (7.3)$$

Eventually all three partials meet at the apex of a completed tetrahedron which has four triangular faces of stacking fault bounded by stair-rod dislocations along the edges.

The reactions pictured above proceed equally well whether the original Frank loop is vacancy or interstitial since the energy considerations are the same in both cases. However this mechanism must involve some migration of point defects within the original loop, since its shape must be transformed into an equilateral triangle. It is conceivable that while this redistribution is possible for a vacancy loop it is not for an interstitial loop, since a different point defect is migrating in these two cases. The fact that the transformation does not occur until a temperature range (250 to 350°C) where limited vacancy self-diffusion becomes possible suggests that in this case the migrating point defect is the vacancy and that the observed tetrahedra must therefore be of the vacancy type.

There is also some experimental evidence in support of the conclusion that interstitial tetrahedra do not grow in irradiated gold. Gold foils bombarded with low-energy inert gas ions show black spot and loop damage which is thought to be entirely interstitial (Venables and Balluffi, 1965; Bowden and Brandon, 1963). On annealing, these clusters disappear and no tetrahedra are formed (Ogilvie, Saunders, and Thomson, 1963). Although it is possible, though not very likely, that no Frank loops occur under an irradiation of this sort, it does indicate that interstitial tetrahedra do not normally occur.

It thus seems possible that the tetrahedra observed in as-irradiated and annealed gold films are vacancy, although a completely unambiguous conclusion is not possible on the basis of present observations.

Interpretation of the Annealing Stages

By implication the upper stage has already been associated with stacking fault tetrahedra. However the temperature range is a good deal below the figure (650°C) which is normally associated with the annealing of tetrahedra in quenched gold. We consider that this is probably a size effect. The tetrahedra observed in quenched gold generally have edge lengths of the order of 100 to 600 Å, while almost all of those observed in irradiated specimens are smaller than this.

The most generally accepted model for the shrinkage of tetrahedra is the emission of vacancies by the propagation of jog lines across a face. The theory of this process has been treated in detail by Kuhlmann-Wilsdorf (1965). Since the activation energy required to form such jog lines is proportional to their length, the predicted annealing rates are faster for small tetrahedra where the jog lines are correspondingly short. This agrees with our qualitative observation that the small tetrahedra do in fact disappear first. Moreover, the results of 30 min isochronal anneals at temperatures of 500°C and above are in good quantitative agreement with the experimental results of Segall, Clarebrough, and Loretto (1966), on the annealing of tetrahedra in quenched gold. Their results, however, predict very little annealing of tetrahedra below 500°C. It is possible that our upper stage annealing observed between 350 and 500°C may be surface activated or be due to mutated tetrahedra. The important point is that the observed annealing below 650°C is not inconsistent with that calculated for *small* tetrahedra.

The gradual decrease in cluster size observed in the lower annealing stage suggests that the process involved is one of thermally activated climb rather than precipitous slip to a surface.

We review briefly the current theories on the shrinkage of dislocation loops by climb. This problem was first treated by Silcox and Whelan (1960) using a model for climb originally proposed by Friedel (1964). When a dislocation line with an edge component climbs normal to its glide plane a continuous layer of point defects must be either supplied from the crystal or created at the line. Unless a very strong external stress is present these defects are supplied by a thermally activated process, and, since the simultaneous emission or absorption of N point defects has a probability factor of the order of $\exp(-NE_f/kT)$, such processes are negligible except when $N = 1$. Also, since the energy of formation, E_f, is generally much less for a vacancy than an interstitial, only vacancies need be considered in most climb processes. Single vacancies are most easily emitted or absorbed in dislocation lines at jogs, so that thermally activated dislocation climb is only possible through the movement of jogs.

Silcox and Whelan have considered the case of prismatic vacancy loops. If a vacancy is absorbed at a jog on such a loop the jog moves in such a direction so as to increase the size of the loop, and, of course, in the opposite direction if a vacancy is emitted. The energy necessary to create and emit a vacancy from a loop is equal to the sum of the formation energy, E_f, and the migration energy, E_m, necessary to move it one step away, less a term Fa^2, which represents the decrease in the self-energy of the loop when one vacancy is emitted. By considering the rates of vacancy emission and absorption, Silcox and Whelan arrived at the formula†:

$$\frac{dr}{dt} = -c_j \, zvr_0 \exp\left(-\frac{E_f + E_m}{kT}\right)\left[\exp\left(\frac{Fa^2}{kT}\right) - \frac{c}{c_0}\right] \qquad (7.4)$$

† The formula actually given by Silcox and Whelan differs in the term Fa^2 which they write as Fr_0^2. We have considered that the effective surface area of one point defect within a loop is dependent on the Burger's vector. The major effect of this assumption is to make the annealing rates of perfect and Frank loops approximately equal.

where: r is the radius of a circular dislocation loop; c_j is the 'concentration' of jogs, normally taken as $\frac{1}{2}$ for a dislocation loop; Z is the co-ordination number for vacancy migration (11); ν is the atomic vibrational frequency ($\sim 10^{13}$ sec^{-1}); r_0 is the interatomic distance (2·88 Å); F is the self-energy force; a^2 is the surface area of one point defect with a loop; c is the atomic vacancy concentration in the immediate vicinity of the loop; and c_0 is the equilibrium vacancy concentration: $c_0 \sim \exp[-(E_m + E_f)/kT]$. The first term in the square brackets in (7.4) is due to the emission of vacancies, and the second is due to absorption.

In the case of a perfect loop, F is only due to the line tension in the loop:

$$F = F_\tau = \frac{Gb}{4\pi(1-\sigma)} \frac{\log_e(r/r_0)}{(r/b)} \tag{7.5}$$

where: G is the shear modulus; σ is Poisson's ratio; and b is the Burger's vector. For a Frank faulted loop:

$$F = F + \gamma \tag{7.6}$$

where γ is the stacking fault energy.

Equation 7.4 may be solved to give the loop radius as a function of annealing time assuming $c/c_0 = 1$. The term $[\exp(Fa^2/kT) - 1]$ may be approximated by $\alpha[b/r]$, where α is a parameter depending on the character of the loop and the temperature. This gives:

$$r(t) = R[1 - (t/\tau)]^{1/2} \tag{7.7}$$

where
$$\tau = \frac{R^2 \exp[(E_m + E_f)/kT]}{2C_j Z r_0 \nu b \alpha} \tag{7.8}$$

The values of α and the lifetimes, τ, for a loop of initial radius, $R = 50$ Å, are listed in Table VI for both perfect and Frank loops. We have taken $G = 2·78 \times 10^{11}$ dyn/cm^2, $\sigma = 0·42$, and $\gamma = 20$ erg/cm^2.† The almost identical values of α and τ for both perfect and faulted loops is a direct consequence of using a term Fa^2 instead of Fr_0^2.

The case of an interstitial loop is somewhat different. Here the probability of vacancy emission from a jog is negligible since this process must involve the creation of a vacancy-interstitial pair with a prohibitively high activation energy (~ 5 eV). The sole process involved is vacancy absorption from the lattice, so that the annealing rate is given by:

$$\frac{dr}{dt} = -c_j Z \nu r_0 \exp[-(E_m + E_f)/kT] c/c_0 \tag{7.9}$$

$$r(t) = R[1 - (t/\tau)] \tag{7.10}$$

† The results are not highly sensitive to the value chosen for γ. F is of the order of 100 erg/cm^2 for the loop radii of interest, and the stacking fault energy in gold certainly has a much lower value.

and
$$\tau = \frac{R.\exp\left[(E_m + E_f)/kT\right]}{c_j vZr_0(c/c_0)} \tag{7.11}$$

The lifetimes, τ, listed in Table VI for $R = 50$ Å apply approximately to both perfect and Frank interstitial loops. There is a slight difference due to the different Burger's vectors involved, but this appears only in the pre-exponential term and, considering the overall approximations, the effect cannot be critical. We have again assumed $c/c_0 = 1$. We note from Table VI that the expected lifetimes for interstitial loops are approximately five times longer than those expected for vacancy loops of the same initial size. It might be hoped that this criterion would be sufficient to distinguish experimentally between the two types.

TABLE VI

Temperature (°C)	Perfect vacancy loops		Frank vacancy loops		Interstitial loops
	α	τ (min)	α	τ (min)	τ (min)
250	69	200	66	260	1600
300	56	7	55	9	50
350	46	0·5	47	0·5	2·5

Experimentally, however, the lifetime of clusters which anneal out in the lower stage is very much more than 5 h at 250°C, of the order of 5 h at 300°C, and less than 10 min at 350°C. These lifetimes are very much greater than the values of τ listed in Table VI for either vacancy or interstitial loops. We shall consider two possible explanations for this large discrepancy.

Firstly, it may well be that the rate of climb given by the Silcox-Whelan model is not strictly correct. In their derivation of equation 7.4 they assume that for loops sufficiently close to a surface all emitted vacancies are removed to the surface and no re-absorption occurs ('emission-limited climb'). However Seidman and Balluffi (1966) argue that the rate of escape of vacancies from an emitting loop must always be limited by the diffusion conditions, and that a localized supersaturation of vacancies is inevitably created in the vincinity of such a loop, thus enhancing the absorption of vacancies. In the 'diffusion-limited climb' model the rate of climb is given by an equation qualitatively similar to equation 7.4, the only difference arising in the pre-exponential term. It is obvious that under normal conditions the emission-limited model gives the maximum shrinkage rate. Seidman and Balluffi (1966) have in fact estimated, on the basis of one specific model, that the rate of climb given by the diffusion-limited model is some ten times slower than that given by the emission-limited model. Increasing the values of τ in Table VI by a factor of 10 would bring them into an order of magnitude agreement with the experimental results. The theory, however, cannot be considered sufficiently accurate to justify distinguishing between vacancy and interstitial loops purely on the basis of their estimated lifetimes.

Another possible explanation for the slow annealing rate is that most of the clusters are Frank dislocation loops which, by their dissociation, retard the propagation of jogs essential for climb. Under these circumstances the moving jogs must now move, at least some of the time, across dissociated segments of the dislocation line. To do this, simple jogs must be extended across the faulted face, and the additional activation energy, ΔE, necessary to extend a jog decreases its rate of propagation by a factor of the order $\exp(-\Delta E/kT)$. Since one step of an extended jog increases the length of the jog line by something of the order of one interatomic distance, ΔE is of the order of E_j, the jog-line energy per atom plane. If $E_j = 0.073$ eV, then at 300°C $\exp(-\Delta E/kT) \sim 1/5$, and the lifetime of a dissociated loop would be of the order of five times that of an undissociated loop. The lifetimes of undissociated Frank loops listed in Table VI are perhaps 1/30 of the experimental lifetimes, so the proposed effect does not yield entirely unreasonable predictions. There is, of course, no inherent reason why both these explanations cannot be valid, and that the climb of dislocation loops is both diffusion-limited and retarded by the dissociation along Frank loops.

We note that the distinction between dissociated and undissociated loops does not apply to interstitial loops. The energy of formation which is released when a vacancy is absorbed at a jog on an interstitial loop is very much more than the energy required to move the jog whether it is extended or not, so that all vacancies reaching a jog on an interstitial loop should be automatically absorbed. The climb rate of an interstitial loop is still given by equation 7.9. If the rate of climb of vacancy loops is indeed reduced by a factor of perhaps 1/30 by their dissociation, then it is possible that interstitial loops would climb at a faster rate. This agrees with the results of Ogilvie *et al.* (1963) who find that what they interpret as interstitial loops in gold anneal out in a temperature range of from 220 to 280°C.

Moreover, dissociation may account for the fact that the first sign of annealing in the lower stage is a gradual shrinkage of the larger loops. This appears to contradict the basic models of climb, since the annealing rate for vacancy loops should be greater for small loops, while for interstitial loops it should be independent of size. Thus for either type of loop the shrinkage of large loops should be accompanied by the rapid disappearance of smaller loops, and this behaviour is predicted by both the diffusion-limited and emission-limited models of climb. Experimentally, this is definitely not observed. If, however, the larger loops are either perfect loops or undissociated Frank loops, their annealing rate is not retarded by dissociation, and they should anneal at a faster rate than smaller, dissociated loops. Perfect loops, unless they transform to Frank loops during annealing, should have relatively short lifetimes. The fact that the total concentration of clusters decreases only slightly on prolonged annealing at 250°C implies, if our assumptions are correct, that the fraction of perfect loops is quite small, while the majority are dissociated Frank loops. This is in basic accord with our previous conclusions.

The other major feature of the lower annealing stage is the formation of tetrahedra. Their growth is easily explained by the Silcox-Hirsch model (1959), outlined above, namely that they grow by the dissociation of a Frank sessile loop onto adjoining {111} planes. Since the concentration of tetrahedra is always about 40% of the original cluster concentration, at least 40% of the clusters are initially Frank loops.

Why do some clusters anneal out in the lower stage while others form tetrahedra? Since the effect is independent of thickness we may exclude the possibility that the fate of a cluster is somehow determined by its distance from the surface. Perfect loops must account for some of the clusters which anneal out, since they cannot dissociate into tetrahedra. However, the results of our experimental observations indicate that less than 25% of all clusters are perfect loops, so they cannot account for the entire effect. Another explanation might be that some interstitial loops are present since, as mentioned above, there is experimental evidence that they anneal out rather than form tetrahedra. It may be shown that the formation of interstitial loops critically depends on the concentration of single interstitials, and that this in turn must be strongly dependent on the flux and film thickness. Since the proportion of clusters which anneal out in the lower stage is independent of the flux and film thickness it appears unlikely that these clusters are interstitial loops.

A better explanation may be that not all Frank loops form tetrahedra. A tetrahedron may only grow from a Frank loop if the dissociated Shockley partials all lie on the same side of the original loop plane. Since the dissociation may occur in six different ways – along one of three $\langle 110 \rangle$ directions and on either side of the loop plane – it is very probable that initially the Shockley partials lie on both sides of the plane. However, the elastic interaction between partial dislocations on opposite sides may be sufficient to draw them all onto the one side. If this is not the case, a tetrahedron can never form, and the loop must anneal out by climb. If the probability of such an event is determined purely by chance, then the ratio of Frank loops which form tetrahedra to those which anneal out is independent of the total loop concentration, in accord with the observed behaviour.

The explanation of the lower annealing stage most consistent with both the growth of tetrahedra and the shrinking of clusters is that all clusters are initially dislocation loops of vacancy character (except of course the tetrahedra formed during the irradiation). Perfect loops anneal out relatively rapidly by climb and are not numerous. Frank loops either dissociate fully to form stacking fault tetrahedra, or partially, whereupon they shrink at a retarded rate. The evidence for such conclusions is somewhat tenuous, but we believe it to be a reasonable representation of what actually occurs.

Conclusion

Electron microscopy of fission fragment irradiated single crystals of gold therefore suggests the following:

(1) The effect of the irradiation is to introduce a random array of defect clusters. Tracks are *not* observed.

(2) Almost all of the visible clusters are Frank faulted dislocation loops ($b = \frac{1}{3}\langle 111 \rangle$), containing 40 to 1100 point defects.

(3) Less than 20% of the clusters can be ascribed to perfect dislocation loops with the Burger's vector in the plane of the film, and less than 5% to perfect loops with an inclined Burger's vector.

(4) Some spherical clusters may be present but at most they comprise less than 33% of the total concentration.

(5) Simple methods do not distinguish between clusters of vacancy or interstitial character.

(6) There are two distinct annealing stages. A lower stage centred in the temperature range between 250 and 350°C is associated with the thermally activated climb of dislocation loops. An upper stage (350 to 600°C) is associated with the annealing of vacancy tetrahedra.

References

ASHBY, M. F., and BROWN, L. M. (1963a), *Phil. Mag.* **8**, 1083.

ASHBY, M. F., and BROWN, L. M. (1963b), *Phil. Mag.* **8**, 1964.

BOWDEN, P. B., and BRANDON, D. G. (1963), *J. Nucl. Mat.* **9**, 348.

FRIEDEL, J. (1964), *Dislocations*, Pergamon Press (London).

HESKETH, P. V., and RICKARDS, G. K. (1966), *Proc. Roy. Soc. A* **289**, 353.

HIRSCH, P. B., HOWIE, A., NICHOLSON, R. B., PASHLEY, D. W., and WHELAN, M. J. (1965), *The Electron Microscopy of Thin Crystals*, Butterworths (London).

KUHLMANN-WILSDORF, D. (1965), *Acta Met.* **13**, 257.

MCINTYRE, K. G., and BROWN, L. M. (1966), *Consultants Symposium on the Nature of Small Defect Clusters*, Harwell, UK.

MCINTYRE, K. G., and BROWN, L. M. (1967), to be published.

OGILVIE, G. J., SAUNDERS, J. V., and THOMSON, A. A. (1963), *J. Phys. Chem. Sol.* **24**, 247.

PASHLEY, D. W. (1959), *Phil. Mag.* **4**, 324.

PASHLEY, D. W., and PRESLAND, A. E. B. (1960), *Proc. Europ. Reg. Conf. Electr. Microsc.*, Delft, p. 391.

RUEDL, E., DELAVIGNETTE, P., and AMELINCKX, S. (1962), *J. Nucl. Mat.* **6**, 46.

RUHLE, M., WILKENS, M., and ESSMAN, U. (1965), *Phys. Stat. Sol.* **11**, 819.

SEGALL, R. L., CLAREBROUGH, L. M., and LORETTO, M. H. (1966), *Phil. Mag.* **14**, 53.

SEIDMAN, D. N., and BALUFFI, R. W. (1966), *Phil. Mag.* **13**, 649.

SIGLER, J. A., and KUHLMANN-WILSDORF, D. (1966), *Consultants Symposium on the Nature of Small Defect Clusters*, Harwell, UK.

SILCOX, J., and HIRSCH, P. B. (1959), *Phil. Mag.* **4**, 72.

SILCOX, J., and WHELAN, M. J. (1960), *Phil. Mag.* **5**, 1.

VAN VLIET, D. (1966), Ph.D. Dissertation, Cambridge.

VAN VLIET, D. and CHADDERTON, L. T. (1967), *Phil. Mag.* **16**, 275.

VENABLES, J. A., and BALLUFFI, R. W. (1965), *Phil. Mag.* **11**, 1021.

CHAPTER EIGHT

Heat-Sensitive Crystals

A large fraction of the energy lost by a fission fragment passing through a solid is expended in electronic excitation and ionization, and eventually appears as heat. This chapter describes the results of an attempt to observe, in as much detail as possible, the damage produced by the fission fragment irradiation of some solids which were likely to be sensitive to this heating effect, and whose crystal lattices would not be expected to recover from such damage. The discussion considers whether the damage can be explained in terms of heat effects alone. Observations of this kind are of interest in connection with the concept of the minimum hot-spot size for the initiation of a thermal explosion (see Bowden and Yoffe, 1958), and with the problem of radiation-induced sensitivity to subsequent thermal decomposition (Young, 1960).

It was necessary that the materials, while being liable to thermal decomposition or evaporation, should withstand the beam of an electron microscope, and that crystals of about 300 Å thickness, i.e. thin enough for direct transmission electron microscopy, should be obtainable. The following substances were chosen: silver cyanamide; the thiocyanates of silver and mercury (II); the picrates of ammonium, potassium, and thallium; ammonium iodate; and ammonium borofluoride. Preliminary observations on tracks in silver cyanamide have been published previously (Bowden and Montagu-Pollock, 1961).

Experimental

Silver cyanamide crystals were grown from solution by a method described elsewhere (Montagu-Pollock, 1962). Silver thiocyanate was prepared from silver nitrate and potassium thiocyanate, and crystals thin enough to give strong optical interference colours were grown by the slow evaporation of ammonia from an ammoniacal solution of the thiocyanate. These thin crystals were found to be stable to light, while the thicker and irregular crystals which were also produced were seen to decompose rapidly. Drops of suspensions of both the cyanamide and the thiocyanate crystals in water were deposited on carbon films supported by electron microscope specimen grids of very fine mesh (500 bars to the inch). Mercuric thiocyanate was prepared by double decomposition. Potassium picrate and ammonium picrate specimens were

112

prepared as follows: a weak solution (2 or 3 %) of the material was made in carefully dried ethanol or methanol, and carbon-coated grids, similar to those described above, were placed on slightly inclined glass slides. Drops of the solution were placed on the grids and allowed to run down the slides. When the alcohol had evaporated, the carbon films were found to be coated with very thin crystalline films of the material. Thallium picrate, ammonium borofluoride, and ammonium iodate crystals were prepared from a weak solution of the material in water. Grids coated with collodion followed by carbon were placed on a metal plate for good thermal conduction, in a vacuum bell-jar. A drop of solution was placed on each grid, and the bell-jar was replaced and evacuated by means of a powerful rotary pump. The water evaporated rapidly and the collodion-carbon films were found to be covered with very thin crystals.

Random fission fragment irradiation was carried out using natural uranium in the manner described in Chapter 3. The irradiation time was in general not greater than 5 min. The crystals were then photographed in the electron microscope (a Siemens Elmiskop I operated at 80 kV) at magnifications of up to 20,000 times. The use of very fine mesh grids was found to prevent decomposition due to the heating effect of the electron beam, provided that the intensity was kept below a certain level. The need to keep the beam below this level, however, meant that the higher magnification ranges could only be used with difficulty.

Results

In some of the materials, clearly defined tracks were visible on the electron micrographs and the number of tracks per unit calculated flux of fission fragments did not appear to vary with the irradiated solid. The micrographs were again consistent with the assumption that where tracks are seen, each track was produced by one fission fragment. Control specimens, irradiated with neutrons and gamma-rays only, never showed tracks.

Silver Cyanamide

After irradiation, continuous tracks of light contrast appeared; the long tracks representing paths of fragments travelling almost in the plane of the crystal, while white dots indicated fragments that penetrated the crystal vertically, or nearly so. The mean track diameter was about 120 Å. The visibility of the tracks decreased with increasing crystal thickness, but thick crystals opaque to the electron beam still showed occasional white dots corresponding to vertically impinging fragments. It appears that with crystal thicknesses of up to $0.1\ \mu$ at least, there is no evidence for variation in track width with crystal thickness, distance from the crystal surface, or angle of track. When silver cyanamide decomposes, silver is formed (Montagu-Pollock, 1962), and electron diffraction patterns confirmed that dark spots in fission fragment irradiated crystals were specks of silver, often at large distances from the tracks. Most of this silver clearly resulted from fission fragment damage: controls irradiated without uranium foil required a dose of neutrons and γ-rays one hundred times as large as that received by fragment irradiated crystals to give comparable amounts of silver.

Silver Thiocyanate

Irradiated silver thiocyanate crystals possessed continuous dark tracks. No clearly defined crystal thickness effects were found, though on occasional micrographs of very thin crystals the tracks tended to be of light contrast.

Mercuric Thiocyanate

Irradiated thin crystals of mercuric thiocyanate showed continuous tracks of light contrast. Continuous tracks in thicker areas of a crystal did not in general tend to be longer, suggesting that continuous track registration did not occur below a certain depth from the crystal surface. Thus all tracks except those making a small angle with the crystal surface gave rise to slightly elongated dots on the micrographs corresponding to points of entry or exit, and the proportion of dots to continuous tracks was therefore considerably higher than for silver cyanamide.

Discontinuous tracks were also often seen in the thicker crystals. A typical track making a small angle with the crystal surface was continuous for a few hundred Ångströms from the point of entry or exit, and then either ceased to register or became discontinuous. The width of all tracks whether continuous, discontinuous, or of dot form (corresponding to tracks of high angle), averaged about 75 Å, and the spacing between the dashes of a discontinuous track appeared to average 250 to 300 Å, and to show no regular variation with distance along the track.

In order to confirm that continuous track registration occurred only near the free crystal surface, an irradiation was carried out on crystals whose upper surface had been coated with a layer of evaporated carbon, similar to the support film covering the lower surface. No tracks, continuous or dashed, were seen on the micrographs, but arrays of dark lines suggesting the introduction of a dislocation network appeared and spots of black/white contrast, of a distribution corresponding to that of the tracks seen in uncoated crystals, were observed.

An attempt was made to find, by gold-palladium shadowing, the depth of regions of tracks that were open to the crystal surface. The distribution of shadowing alloy on the crystal surface was not definite enough to give unambiguous micrographs, though a few of the dots and continuous tracks were clearly seen to have been shadowed, while discontinous tracks appeared to be identical to those in unshadowed crystals.

Ammonium Picrate

There was no indication of track formation. Very small and numerous white dots could not definitely be attributed to fission fragment damage, since they were also seen in unirradiated crystals, and were probably produced by the electron beam.

Potassium Picrate

Fission fragments produced no visible tracks in potassium picrate crystals. However, small widely spaced dots of black/white contrast appeared. These were never seen in unirradiated crystals.

Thallium Picrate

Continuous tracks of light contrast were visible in thallium picrate crystals. The track width decreased considerably with increasing crystal thickness. Crystals of still greater thickness showed spots alone, corresponding to the points of entry or exit of the fission fragments.

Ammonium Iodate

No damage was visible on micrographs of irradiated ammonium iodate crystals. The Bragg contours appeared to be irregular and broken into specks in some areas, in both unirradiated and irradiated crystals. This was probably an effect of the electron beam (see, for example, Tubbs and Forty, 1962).

Ammonium Borofluoride

Contrast effects characteristic of alined dislocation loops were seen on micrographs of irradiated ammonium borofluoride crystals. No tracks were visible.

Discussion

A striking characteristic of the results was the great variation, from one substance to another, in the type and amount of damage produced by the fission fragment irradiation. Any explanation of the observed tracks of individual fragments must account for continuous and discontinuous tracks, of dark or light contrast compared with neighbouring areas of the crystal, and for track diameters varying over a wide range.

Light and Dark Tracks

It is assumed that tracks of light contrast represent volumes from which material has escaped, whether they are below the crystal surface or form a trough along it. The dimensions of such tracks can therefore indicate with some precision the volume of missing material. Features arising from diffraction contrast, however, show fluctuations in brightness or are simply dark. In Chapter 5 we discussed examples of fission fragment tracks of this type in molybdenum disulphide, and the dark tracks in silver thiocyanate similarly represent regions of strained or disordered crystal (whether decomposed or not) which give rise to diffraction contrast. With this type of track, the amount of material responsible for the effect cannot be reliably estimated.

The Effect of Heat

With the exception of the features observed on micrographs of irradiated ammonium borofluoride and potassium picrate, and already discussed, it is proposed that all the observations can be explained in terms of a heating effect arising from the transfer of energy from the fragment to electrons in the irradiated crystals, which is the predominant means of energy transfer. Experimental values of ranges of uranium fission fragments are few (see, for example, Noggle and Stiegler, 1960) but indicate that in substances of low and medium atomic number, an average fragment of energy 80 MeV and velocity $1 \cdot 2 \times 10^9$ cm/sec loses $c. 10^3$ eV/Å of track. Calculations of the 'electronic' energy loss by the Bethe/Bloch formula (see Bethe and Ashkin, 1953, and Chapter 2) give values of around five times this figure, but confirm that for substances of low and medium atomic number the energy losses per Ångström of track vary by only a small factor. In this chapter we will adopt the figure of 10^3 eV/Å for all the substances, since the measure of agreement between experiment and theory does not justify the use of more precise values.

If the whole of this energy were to be effective in producing decomposition, each fragment would give a cylinder of damage at least 200 Å wide (with or without melting, sublimation, and recrystallization) penetrating the crystal. The visible amount of damage was of much smaller extent, no tracks being seen, for example, in potassium picrate. The possibility of decomposition followed by recombination is most unlikely to account for the discrepancy, at least as regards the exothermic substances (cyanamides, thiocyanates, and picrates, the latter being highly explosive), while even in endothermic materials the dense concentration of energy reduces the possible extent of recombination (Bertocci, Jacobi, and Walton, 1961). Furthermore, it will be shown that, on the basis of a simple model of thermal damage, variation in thermal conductivity is not sufficient to account for the large variation, from one substance to another, in the amount of observed damage; and that the volume within which the energy is released as heat is the critical parameter. The observations enable rough limits to be calculated for this volume.

The model takes the electronic energy given to the crystal by the fission fragment to appear eventually as a quantity of heat Q per Ångström of track, uniformly and instantaneously distributed within a cylinder of radius r_0

described about the track. Discussion of this liberation of heat is deferred to Chapter 12. The laws of thermal conductivity are taken to hold: computer calculations by Vineyard (1961) have shown that for atomic collision processes of even shorter times than those involved here, macroscopic thermal concepts still have some validity with thermal conductivities higher than macroscopic values by a factor of about ten. Heat losses by radiation are neglected. The following symbols are used:

$T(r,t)$ temperature at a distance r from the track at time t
Q, r_0 (defined above)
k Boltzmann's constant
c, ρ, λ specific heat, density, thermal conductivity
$D = \lambda/c\rho$ thermal diffusivity
ν_0 lattice vibration frequency

Calculation of the amount of damage is simplified by the following approximations:

(1) A molecule is assumed to be 'damaged' (decomposed or sublimed) if it is thermally activated at any time to an energy E.

(2) Within a certain distance from the track, damage is complete. At this distance (r_1) a molecule is excited just once to the activation energy and r_1 is given by:

$$1 = \nu_0 \int_{\text{all } t} \exp\left[-\frac{E}{kT(r_1, t)}\right] dt \qquad (8.1)$$

provided that this gives $r_1 \geqslant r_0$; otherwise $r_1 = 0$. At greater distances, consumption of reactant is neglected, the decomposition being taken as a zero-order reaction, so that beyond $r = r_1$ the number of damaged molecules is proportional to r_2^2 where:

$$r_2^2 = \nu_0 \int_{r_1}^{\infty} \int_{\text{all } t} 2\pi r \exp\left[-\frac{E}{kT(r, t)}\right] dr\, dt \qquad (8.2)$$

(3) It is assumed that the undamaged molecules will then recrystallize, leaving a cylinder of damage equal in volume to the sum of the original volumes of all the damaged molecules. Its radius is thus given by R where:

$$R^2 = r_1^2 + r_2^2 \qquad (8.3)$$

This assumption of a sharply defined cylinder of damage is consistent with the observed appearance of a fission fragment track, whose radius then corresponds to R.

(4) The effect of any heat liberated by exothermic decomposition, in both parts, is neglected in comparison with Q. This approximation is expected to be fair for situations where the amount of decomposition is small.

9

Exact evaluation of these integrals is made difficult by the form of $T(r,t)$, which for the simple case $r_0 = 0$ is given (Chapter 12) by:

$$T(r,t) = \frac{Q}{4\pi\lambda t} \exp\left(-\frac{r^2}{4Dt}\right) \tag{8.4}$$

(see, for example, Dienes and Vineyard, 1957). However, a good approximation to this initially infinitesimally narrow thermal spike is:

$$\begin{aligned} \text{for } r < \sqrt{(Dt)}, \quad &T = Q/8\sqrt{\pi\lambda t} \\ \text{for } r > \sqrt{(4Dt)}, \quad &T = 0 \end{aligned} \tag{8.5}$$

Our thermal spike has initially a finite radius r_0, but the same approximation will serve provided that no events before a time t_0 are considered, where:

$$t_0 = r_0^2/4D \tag{8.6}$$

R can now be calculated. If $r_1 \geqslant r_0$, a molecule at $r = r_1$ is not heated until time $t = r_1^2/4D$, and the amount of damage is the same as if r_0 and t_0 were zero. We then have, from (8.1) and (8.5):

$$r_1 = v_0 \int_{t=r_1^2/4D}^{\infty} \exp\left(-\frac{E}{Q}\frac{8\sqrt{\pi\lambda t}}{k}\right) dt \tag{8.7}$$

giving:

$$r_1^2 = \frac{k}{2\sqrt{\pi}c\rho}\frac{Q}{E}\ln\left(\frac{v_0 k}{8\sqrt{\pi\lambda}}\frac{Q}{E}\right) \tag{8.8}$$

and from (8.2) and (8.5):

$$r_2^2 = v_0 \int_{t=r_1^2/4D}^{\infty} (4Dt - r_1^2)\exp\left(-\frac{E}{Q}\frac{8\sqrt{\pi\lambda t}}{k}\right) dt \tag{8.9}$$

which from (8.8) gives:

$$r_2^2 = \frac{k}{2\sqrt{\pi}c\rho}\frac{Q}{E} \tag{8.10}$$

so that from (8.3):

$$R^2 = \frac{k}{2\sqrt{\pi}c\rho}\frac{Q}{E}\left[\ln\left(\frac{v_0 k}{8\sqrt{\pi\lambda}}\frac{Q}{E}\right) + 1\right] \tag{8.11}$$

With the given temperature distribution (8.5, 8.6), all the molecules within $r = r_0$ have the same thermal history, so that the only alternative to $r_1 \geqslant r_0$ is $r_1 = 0$. In this case we have, from (8.2) and (8.3):

$$R^2 = r_2^2 = v_0 \int_{t=r_0^2/4D}^{\infty} 4Dt \exp\left(-\frac{E}{Q}\frac{8\sqrt{\pi\lambda t}}{k}\right) dt \tag{8.12}$$

giving:

$$R^2 = \frac{\nu_0 k}{8\sqrt{\pi\lambda}} \frac{Q}{E}\left(r_0^2 + \frac{k}{2\sqrt{\pi c\rho}} \frac{Q}{E}\right)\exp\left(\frac{-r_0^2}{(k/2\sqrt{\pi c\rho})(Q/E)}\right) \qquad (8.13)$$

Comparison of Observed and Calculated Track Widths

The presence of silver on irradiated silver cyanamide crystals implies that some decomposition has taken place. On the assumption that the observed tracks represent tunnels of decomposition, either equation 8.11 or 8.13 is successful in giving a value of R equal to the observed track radius of 60 Å. The following are the values of the thermal parameters used:

$$
\begin{array}{ll}
c & 10^{-1} \text{ cal/g} \\
\rho & 5 \text{ g/cm}^3 \\
\nu_0 & 10^{13.5} \text{ (see Deb and Yoffe, 1959)} \\
E & 2 \text{ eV (see Montagu-Pollock, 1962)} \\
Q & 10^{11} \text{ eV/cm (see above).}
\end{array}
$$

Apart from r_0, λ is the most uncertain of the parameters, and should probably be higher than the macroscopic value. Table VII lists pairs of values of λ and

TABLE VII

λ (cal/deg C sec cm)	Condition	r_0^2 (cm²)
10^{-2}	$r_1 \geqslant r_0$	3×10^{-13}
	$r_1 = 0$	4×10^{-13}
10^{-3}	$r_1 = 0$	5×10^{-13}
10^{-4}	$r_1 = 0$	10^{-12}

r_0^2 which give the correct value of R. As has been discussed, the thermal conductivity may be higher than the low-temperature value though it probably lies within the above range of values. It is likely therefore that, in silver cyanamide, r_0^2 is not greater than a small multiple of 10^{-13}.

Let us now compare the observed track radii in the two materials of greatest and least sensitivity to irradiation, i.e. silver cyanamide and potassium picrate, whose properties will be denoted by suffixes a and b respectively. Silver cyanamide shows tracks of diameter 120 Å, while no tracks are detectable in irradiated potassium picrate. On the basis of our model we may therefore say that $R_a^2/R_b^2 > 100$. Equation 8.11 will not fit the potassium picrate data unless c, E, Q, ρ, ν_0, or λ differs from the values listed for silver cyanamide by a very large factor indeed; in fact E_b is probably less than E_a, as the explosion temperatures are: silver cyanamide, 470°C; potassium picrate, 310°C. The difference between R_a and R_b is therefore likely to arise mainly from differing values of the exponential term:

$$\exp \left(\frac{r_0^2}{(k/2\sqrt{\pi c \rho})(Q/E)} \right)$$

in equation 8.13. This does not depend on the thermal conductivity. It is evident that a difference in r_0 alone is sufficient to account for the large R_a^2/R_b^2 ratio. Within the range of λ considered, and provided that the other thermal parameters are given the same values as above for both substances, the ratio calculated from equation 8.13 agrees with that observed if, for example, $r_{0,a} = 4\cdot0 \times 10^{-13}$ and $r_{0,b} = 4\cdot5 \times 10^{-13}$, or if $r_{0,a} = 1\cdot00 \times 10^{-12}$ and $r_{0,b} = 1\cdot02 \times 10^{-12}$. These figures show that a very small increase in r_0 reduces very greatly the amount of damage. A more sophisticated model, using accurate values of thermal parameters, would be likely to give different values of r_0. However, the approximate formulae are enough to show that the amount of damage produced by a fission fragment depends very critically on the volume within which its energy appears as heat in the crystal.

The absence of tracks in irradiated crystals of ammonium picrate, iodate, and borofluoride suggests that in these materials, as in potassium picrate, r_0 is large and that the heat released by the fragment is not concentrated enough to produce visible decomposition. It is possible, however, that, in the iodate and borofluoride, decomposition is followed by recombination. Point defect mobility may also be an important factor (Chapter 9).

In silver thiocyanate, r_0 is not small enough to give the wide cylinder of decomposition seen in silver cyanamide, but is unlikely to be greater than 100 Å, since tracks are visible. For very thin crystals the change to light contrast may arise from the higher heat concentration; if the crystal thickness is less than r_0, higher temperatures are expected, and material may sublime. This thickness effect appears in thallium picrate also, where the tracks are much wider in very thin crystals.

Discontinuous Tracks of Light Contrast

Tracks of this type, appearing in the irradiated mercuric thiocyanate crystals, were also seen in molybdenum trioxide (Chapter 5). Both these materials sublime (below their decomposition temperatures) at temperatures of a few hundred degrees, and it is proposed that, provided r_0 is small enough, fission fragment tracks in materials with this property are of light contrast, can vary in diameter with the distance below the crystal surface, and can be discontinuous. For such materials E (in this case the vaporization energy per molecule) has a value around that used in the calculation above (2 eV), and, on the basis of the same model, r_0 for mercuric thiocyanate is likely to lie between the values given for silver cyanamide and potassium picrate, the mean track width being intermediate.

Near a free surface, most of the molecules reaching the energy E will vaporize and escape: this corresponds to the continuous part of low-angle tracks, and to slightly elongated dots representing steeper tracks. At somewhat greater depths, the pressure is higher and only a proportion of the activated molecules will escape. The remainder will cool and recrystallize before being able to do so, and the dashed appearance of the tracks shows that the material recrystallizes non-uniformly, and generally forms a single crystal with the surrounding material. At still greater depths, or in crystals coated on both sides with a carbon layer, no material can escape: recrystallization is complete and the particle does not register.

Summary of Conditions for Track Formation

The thermal mechanism of fission damage in these materials appears to account for the observations made, and it has not been found necessary to invoke other mechanisms, such as disruption of the material along the track by electrostatic repulsion (see Stiegler and Noggle, 1960), or direct decomposition by the electrons liberated by the fission fragment. If the temperature profile due to electronic excitation is broad and flat, then the thermal effect will not necessarily permanently damage the lattice. In such circumstances the concept of the displacement spike and the mobility of point defects become important, and both have to be considered in discussing the nature of particle tracks. For track formation by thermal means it is first necessary that the electrons should lose their energy within a small volume, probably of radius less than $c.$ 100 Å. The energy is then released within this volume as heat. Then, if the activation energy is low, decomposition or vaporization may take place. If decomposition releases further heat, the tracks will be wider and the crystal is unlikely to recover from the damage. The track will be dark, unless material escapes when it will be of light contrast; this will be most likely if the decomposition products include a gas. With silver cyanamide, the high surface mobility of the solid decomposition product, silver (Montagu-Pollock, 1962), favours its escape out of the track. When vaporization takes place rather than decomposition, material can escape from regions of track near the surface only: at distances greater than a few tens of Ångströms from the surface, not all the evaporated material can escape, and the track becomes discontinuous through recrystallization. Dark tracks may also indicate melting followed by discordant recrystallization, as stated by Walker and Price (1962) for damage in mica, though Bonfiglioni, Ferro, and Mojoni (1961) attribute the effect to decomposition. Severe damage, even for large values of r_0, would be expected in materials capable of direct decomposition by electrons. Such materials, however, may not be examined by direct electron microscopy.

Possibility of Subsequent Damage and Explosion

Sole (unpublished work) has found that the thermal decomposition of silver cyanamide is much more rapid if the crystals have previously been irradiated with fission fragments. He attributes this in part to the large surface area of crystals containing hollow cylinders. It is interesting to note that Walker and Price (1962) have found that etching or phase transformation in mica can take place preferentially at fission fragment tracks. Self-etching effects have also been seen by the authors in stored crystals of molybdenum trioxide and lead iodide.

These examples illustrate that fission fragments tracks can assist subsequent damage to a crystal. However, the heat released during the formation of a fission fragment track in an exothermic crystal is not enough for the damage to grow to explosion. The observed diameters of the tracks are in every case smaller than previous estimates of limiting hot-spot sizes (10^3 to 10^5 Å) for a number of explosives (Bowden and Singh, 1954). This work therefore supports their conclusion that explosion should not be induced by a fission single fragment.

Conclusion

It has been shown that uranium fission fragments produce tracks of different types in single crystals of several heat-sensitive materials, while in others no linear damage is visible. A simple thermal mechanism of damage is put forward, based on the loss of energy of the fission fragment to electrons in the irradiated crystal. Differences in the amount of damage are attributed largely to the variation, from one material to another, in the range of these electrons. This mechanism is to be contrasted with that which operates in more stable crystals where a considerable amount of the permanent damage may be ascribed to displacement cascades.

References

BERTOCCI, U., JACOBI, R. B., and WALTON, G. N. (1961), *Chemical Effects of Nuclear Transformations*, International Atomic Energy Authority (Vienna).

BETHE, H. A., and ASHKIN, J. (1953), *Experimental Nuclear Physics* (ed. Segre), **1**, 166, Chapman and Hall (London).

BONFIGLIONI, G., FERRO, A., and MOJONI, A. (1961), *J. Appl. Phys.* **32**, 2499.

BOWDEN, F. P., and MONTAGU-POLLOCK, H. M. (1961), *Nature*, Lond. **191**, 556.

BOWDEN, F. P., and SINGH, K. (1954), *Proc. Roy. Soc. A* **227**, 22.

BOWDEN, F. P., and YOFFE, A. D. (1958), *Fast Reactions in Solids*, Butterworths (London).

DEB, S. K., and YOFFE, A. D. (1959), *Trans. Faraday Soc.* **55**, 106.

DIENES, G. J., and VINEYARD, C. H. (1957), *Radiation Effects in Solids*, Interscience (New York).

MONTAGU-POLLOCK, H. M. (1962), *Proc. Roy. Soc. A* **269**, 219.

NOGGLE, T. S., and STIEGLER, J. O. (1960), *J. Appl. Phys.* **33**, 1726.

PRICE, P. B., and WALKER, R. M. (1962), U.S. Gen. Elect. Com. Rep. 62-RL-2969 M, 3053 M.

STIEGLER, J. O., and NOGGLE, T. S. (1962), *J. Appl. Phys.* **33**, 1894.

TUBBS, M. R., and FORTY, A. J. (1962), *Phil. Mag.* **7**, 709.

VINEYARD, G. H. (1961), *Disc. Faraday Soc.* No. 31, 7.

WALKER, R. M., and PRICE, P. B. (1962), U.S. Gen. Elect. Co. Rep. 62-RL-3054; *J. Appl. Phys.* **33**, 2625.

YOUNG, D. A. (1960), *J. Chem. Soc.*, p. 4533.

CHAPTER NINE

The Nature of Damage in Lead Iodide

The extent of the damage which is produced by uranium fission fragments in crystal lattices evidently depends on the nature and stability of the bombarded specimens. What other factors are influential in determining the damage? In a series of papers, Forty (1960, 1961a, b) and Dawood and Forty (1962) have described the results of experiments in the electron microscope in which thin crystals of lead iodide (PbI_2) were both electron-irradiated and heated. In each case a solid state reaction of the following kind:

$$\text{solid A} \to \text{solid B} + \text{gas C}$$

was initiated and metallic lead and iodine gas were liberated. It proved possible, moreover, to separate the decomposition into several stages, to follow the formation and motion of lattice defects in some detail, and to determine the sites for preferred deposition of metallic lead. Horne, Ottewill, and McMorris (1961) have made a similar study. In this chapter we describe the results of electron microscope observations of fission fragment irradiated lead iodide (Chadderton and Torrens, 1966) and concern ourselves with the final arrangement of lattice defects rather than with the damaging mechanism itself. Observations of irradiated cadmium iodide are also reported.

Lead iodide has a hexagonal crystal structure with the cadmium iodide or cadmium hydroxide grouping. The space group is D_{3d}^3 and there is a single molecule in the unit cell ($a_0 = b_0 = 4.54$ Å, $c_0 = 6.86$ Å). The lattice consists of sandwiches of planes parallel to the basal (0001) plane, and each 'sandwich' is built up of a plane of lead atoms between two planes of iodine. The space between the sandwiches is relatively large, being 3·43 Å, compared to a distance of 1·71 Å between the planes in each sandwich. The lead–iodine and iodine–iodine distances are closely similar to the sums of the equivalent ionic radii and from this standpoint the crystal may be regarded as being constructed of hexagonally close-packed iodine ions. The smaller lead ions rest at the centres of one-half of the octahedra which are defined by alternate layers of iodine. The crystals generally grow in the form of hexagonal platelets with a basal (0001) plane and the relatively large iodine–iodine distance between molecular sandwiches permits both easy cleavage and glide on this plane.

124

Experimental

Crystals of lead iodide which were thin enough for transmission experiments in the electron microscope were prepared by a number of different methods. It was found possible to obtain reasonably thin specimens by the addition of a few drops of lead nitrate solution to a large volume of potassium iodide solution. In general, however, the external morphology of such crystals was rather poor and the dislocation density was very high. The method which has been described by Forty (1960a), in which a solution in alcohol was saturated at about 100°C in a water bath and allowed to cool slowly, was found to be a little better, though in this case many of the crystals so formed were excessively thick. Therefore the crystals were most often prepared by a modification of the method given by Hibi and Yada (1960). A hot solution of recrystallized lead iodide was made in water and single drops of the solution were placed on the specimen support films. The specimen grids were then placed beneath a vacuum bell-jar of small volume and the system was quickly evacuated. The rapid motion of the solvent across the support film deposited perfect hexagons of lead iodide and it proved to be a simple matter to vary the thickness of these crystals by varying the concentration of the initial solution and the rate of evaporation. Crystals with thicknesses ranging between 30 and 2000 Å were prepared in this manner. The thinnest crystals were quite free from dislocations though there were often crystallographic holes present. Long dislocations and networks of partials were observed in the thicker specimens. When dislocation-free specimens were required for examination these were obtained by sublimation of the raw material in a cylindrical furnace at 360°C, and in a slow stream of argon or nitrogen. Most of the crystals that were grown by this technique were prohibitively thick, but when thin crystals were obtained they were evidently quite perfect. No dislocations could be made visible by diffraction contrast and there were very few steps on the (0001) face.

Irradiation of the crystals with fission fragments was carried out in the manner described previously. Specimens were examined in the Siemens Elmiskop I (at 80 kV) and AEI EM 6 (at 75 kV) instruments, and stationary images were recorded on Ilford Special Contrasty Lantern plates at magnifications of up to 120,000 times. Moving picture sequences were taken on Tri-X reversal film with a Kodak (aperture $f/0.95$) 16 mm camera by directly photographing the screen image through the observation port of the microscope. This permitted a more leisurely study of the dynamics of the complicated defect movements which were often observed (Chadderton, 1962a).

Damage

Throughout the course of the experimental work it was noted that the damage observed in fission fragment irradiated crystals of lead iodide by means of electron microscopy was very dependent on the thickness of the specimen.

Although experimentally the picture was much less well defined, it was possible to a first approximation to make a distinction between the fragment tracks which were observed in thin crystals and the dislocation loops visible in thicker ones.

Tracks

In crystals of thickness less than about 400 Å, fission fragment damage took the form of distinct tracks whose nature was sensitively dependent on the

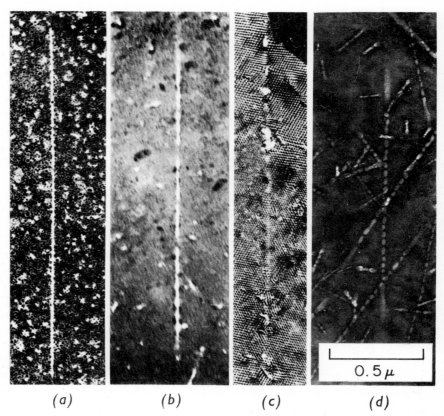

$$(a) \qquad (b) \qquad (c) \qquad (d)$$

Figure 56. Fission fragment tracks in lead iodide may be either continuous, (a), or intermittent, (b). In overlapping crystals, moiré patterns often show regions of strain in the lattice around holes along tracks (c). Tracks in molybdenum trioxide are shown for comparison (d).

thickness of the crystal. In the thinnest crystals (less than 100 Å thick) the tracks were generally of 'white' contrast on a dark ground, due to the removal of material along the path of the fragment. These elongated holes were continuous, or intermittent (Figures 56(a), (b)) with regions where no damage was visible between the segments of a track. They had diameters of up to 150 Å,

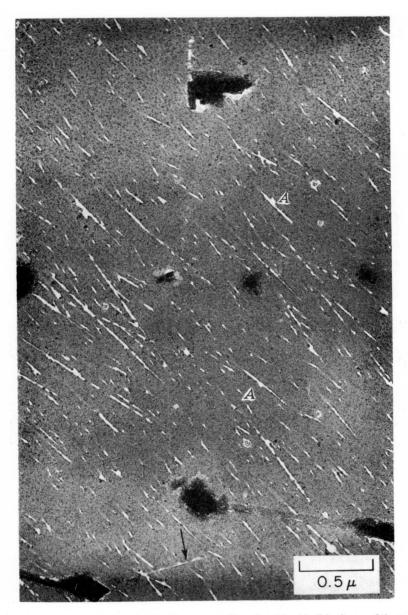

Figure 57. Electron micrograph of a thin crystal of irradiated lead iodide. Some of the tracks have 'spikes' at points where the width is increased by a factor of two or more (points A). (The arrowed track is thinner than those produced by the fission fragments, and is outside the angle of collimation.)

though the diameter was evidently dependent on the energy, charge, and mass of the impinging particle. Often 'spikes', that is small areas where the width was much greater than the average, appeared on tracks (Figure 57). When the tracks were intermittent, the separation between successive segments did not appear to have any regularity, as has been observed in molybdenum disulphide (Chadderton, 1962b), and in some crystals of molybdenum trioxide (Chapter 5). White tracks occasionally could be seen in thin overlapping crystals, and then moiré patterns were resolved in the region of the track (Figure 56(c)).

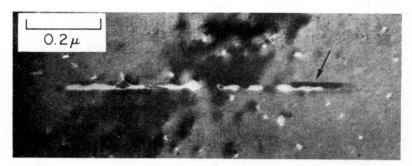

Figure 58. White fission fragment track in lead iodide, along which lead, which was released by the decomposition of the crystal along the path of the particle, has recrystallized (arrow).

The lead that was released by the decomposition of the crystal in the neighbourhood of the path of the particle often crystallized along the track and was oriented with respect to the matrix. Figure 58 shows an area of track containing recrystallized lead (arrow). The electron diffraction pattern from this area revealed face-centred cubic lead in the (111) orientation, superimposed on the hexagonal lead iodide pattern, as was the case with crystals heated in the electron beam (Forty, 1961b). Observation of the crystals with a high-intensity beam indicated preferential decomposition of lead, oriented as above, on the walls and on the environs of the tracks. Extra spots in the diffraction pattern, due to the presence of the lead, were caused by diffraction from planes parallel to the $\{1\bar{2}10\}$ lead iodide set, but of closer spacing ($\sim 1 \cdot 75$ Å as compared to the standard $\{1\bar{2}10\}$ spacing of $2 \cdot 27$ Å). The $\{220\}$ planes of lead have a spacing which is in close agreement with this value. Closer examination of the pattern revealed faint $\{420\}$ spots. The most probable explanation of these is that the lead grows epitaxially in the (111) orientation with its $\{220\}$ planes parallel to the $\{1\bar{2}10\}$ planes of PbI_2, in the form of a large number of small crystals in fibre orientation, rotating about an axis which coincides with one of the principal basal plane directions of PbI_2 (e.g. $\langle 10\bar{1}0 \rangle$).

Large angle deflections of the fragments by close collisions with crystal nuclei were a common occurrence (Figure 59(a)). In Figure 57 the arrowed track is thinner than most fission fragment tracks, and its direction is outside the angle of collimation. It is therefore thought to be a knocked-on atom, probably from a thicker crystal. In crystals less than 50 Å thick, a zone of partial decomposition often surrounded the region of complete decomposition which characterized the track itself (Figure 59(b)).

As the thickness of the crystal was increased to about 250 Å the nature of

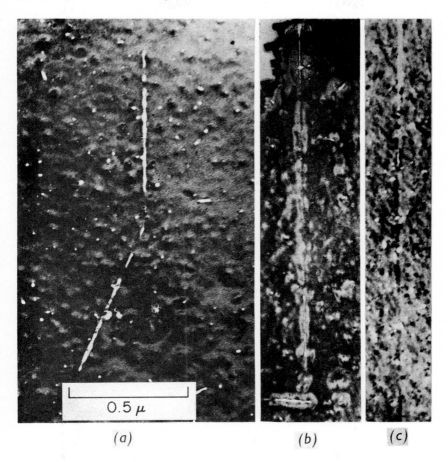

(a)   *(b)*   *(c)*

Figure 59. Further types of fission fragment track in lead iodide include large angle deflections (a) and in crystals less than 50 Å thick, a zone of partial decomposition which surrounds the region of complete decomposition which characterizes the track itself (b). In thicker crystals the tracks become black as diffraction contrast begins to operate (c).

the track changed, diffraction contrast began to operate, and some parts of the track became black (Figure 59(c)). The sensitive dependence of the detailed structure of the black tracks on the diffraction conditions was easily demonstrated by tilting the crystal specimen, so that different sets of atomic planes reflected the electron wave according to Bragg's law. The image of the track was seen to undergo changes which were in general similar to those which may be seen at dislocations (Hirsch, Howie, and Whelan, 1960). These and other contrast phenomena, including tracks exhibiting oscillatory and intermittent contrast, have been discussed in Chapter 6.

In specimens which were placed against uranium foil with a thickness in excess of 20 μ, many unusual tracks were observed (Figure 60). These possessed branches very similar to those reported by other workers (Brostrøm, Bøggild,

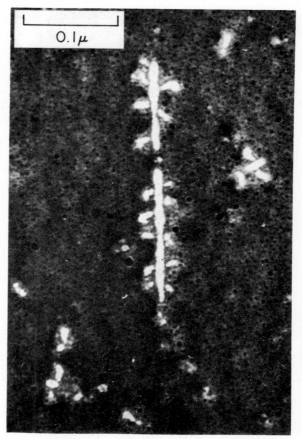

Figure 60. Electron micrograph of fission fragment track in lead iodide. There is a characteristic branching.

and Lauritsen, 1940) in cloud chamber photographs of fission particle tracks. Their occurrence was most marked in the thinnest crystals.

Dislocation Loops

As the thickness of the crystal was further increased, to greater than about 450 Å, tracks were no longer visible, and damage took the form of a random array of dislocation loops which had diameters of up to 250 Å and which were inclined to the surface of the specimen (Figure 61). Only very occasionally were both tracks and dislocations seen in the same crystal, and this was mostly in the neighbourhood of strong extinction contours (Figure 62).

Some of the dislocation loops which were introduced by fission fragment bombardment of relatively thick crystals were revealed by a uniform closed

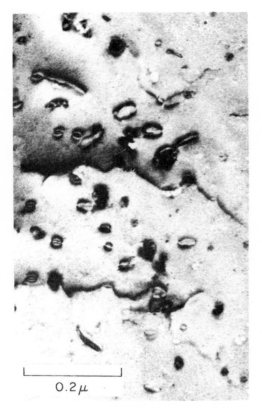

0.2 μ

Figure 61. Dislocation loops are distributed randomly in a thick crystal of lead iodide following fission fragment irradiation.

line of dark diffraction contrast. These were therefore parallel to the surface of the specimen and on (0001) planes. The majority, however, exhibited the well-known elliptical contrast and so were inclined to the basal (0001) plane. A detailed examination of bright-field micrographs and their associated diffraction patterns showed that the major axes of the dark contrast ellipses were parallel to the [11$\bar{2}$0], [$\bar{2}$110], and [1$\bar{2}$10] directions. The crystallographic indices of the loop planes were determined by measurement of ellipse eccentricity on micrographs where two reflection vectors were operating, and where, therefore, the dark contrast loops were double in character, the actual

Figure 62. Electron micrograph showing the comparatively rare occurrence of tracks and dislocation loops in the same area of crystal. A strong extinction contour is necessary in order to resolve the loops.

position of the loop lying between the two electron microscope images. The use of a stereographic projection in the analysis of these data positively established, within the limits of experimental error, that the loops could only lie on members of the $\{10\bar{1}1\}$ family of planes and at an angle of approximately $60°$ to the crystal surface. Two types of plane satisfy this condition. There are the familiar corrugated $(10\bar{1}1)$ type planes with a molecular composition, which are separated by a distance comparable with that between the basal (0001) planes. The $(\bar{1}011)$ type planes, however, are of alternating atomic constitution and are less well defined.

The Burger's vectors of the dislocation loops were investigated by examinations of the contrast in several reflections and by finding the reflections for which the image disappeared. A typical area showing a system of dislocation loops and the appropriate diffraction pattern is shown in Figure 63(a). The diffraction pattern is oriented properly with respect to the micrographs and all electron-optical image rotations have been taken into account. Since the diffraction spots are disposed with hexagonal symmetry and similar intensities about the zero-order spot, it may be concluded that the diffraction vectors from the operating sets of planes (in particular the $\{11\bar{2}0\}$ planes) are excited to the same extent. When the crystal was tilted, however, on the goniometer stage of the electron microscope, it proved possible to operate single reflections to a greater extent than the remainder, and therefore to choose particular sets of planes for preferred diffraction. In Figure 63(b), the $(0\bar{3}30)$ planes are diffracting strongly, and although there is a certain amount of residual contrast at the sites of the dislocation loops due to the other weak reflections (an approximation to the two-beam dynamical theory of electron diffraction is more difficult to achieve in ionic and covalent solids than it is in metals), it may be concluded that the diffraction conditions satisfy the extinction criterion (i.e. $\mathbf{g} \cdot \mathbf{b} = 0$, where $\mathbf{g}$ is the diffraction vector, and $\mathbf{b}$ the Burger's vector). This suggests that in this particular case the loop Burger's vector might well be parallel to $[\bar{2}110]$. Further evidence for this stems from the experimental fact that, under the influence of the electron beam, the loops were found to undergo prismatic glide in this direction.

When the dislocation loops were examined in the electron microscope, they were also observed to move and change in size under the influence of the electron beam. Even at a low beam intensity the smallest loops, which were visible as small black dots, vibrated rapidly. On gradually increasing the intensity of the electron beam, the larger loops also began to move. Figures 64 and 65 show sequences of micrographs, of an area of fragment bombarded crystal containing dislocation loops, taken between incremental increases of the beam intensity. Loops A and B in Figure 64 were subject to a rapid vibration of amplitude of about half a loop diameter, and a slow simultaneous gliding and climbing motion brought them closer together until they finally co-alesced. Occasionally a pair of dislocation loops, for example loops C and D in Figures 65(a), (b), were observed to annihilate each other, in a swift single

10

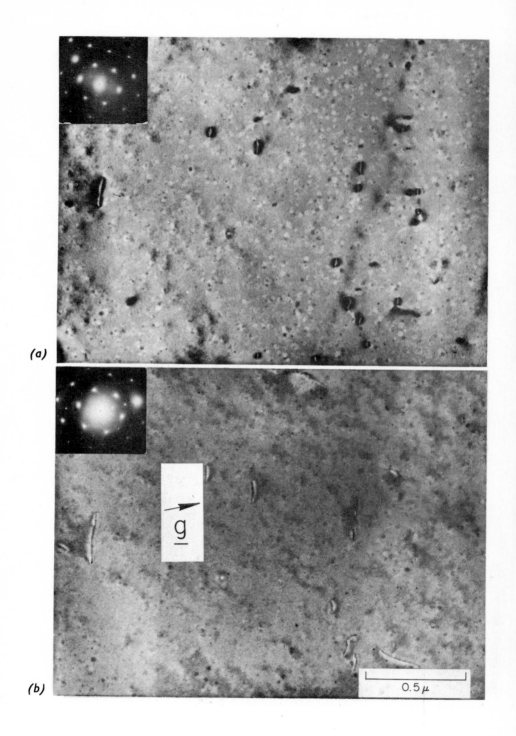

(a)

(b)

g

0.5 μ

movement which was not preceded by either 'slip' or vibration. Further incremental increases in the electron beam intensity increased the rate of growth of the larger dislocation loops, while at the same time many of the smaller loops began to shrink in a fashion very similar to that reported by previous workers (Silcox and Whelan, 1960; Price, 1961) and the shrinkage proceeded faster for loops of smaller diameter. Thus the loop marked E in Figure 65(c) has disappeared completely in Figure 65(d), while loops marked F have experienced only small decreases in diameter.

Under the influence of very high beam currents the crystal quickly became a tangled mass of dislocation lines and dislocation dipoles. Decomposition followed and iodine gas and metallic lead were liberated.

Damage to Cadmium Iodide

Cadmium iodide, like the lead compound, has a hexagonal crystal structure. The interatomic distances are very similar, but the stacking arrangement of the molecular sandwiches is somewhat different. The principal difference in the lattice, however, is evidently that of the atomic masses of the metal atoms (112·4 amu for cadmium and 207·2 amu for lead).

In the thinnest crystals of cadmium iodide, following bombardment with fission fragments, white tracks were again observed (Torrens, 1964). These were similar to those present in irradiated lead iodide, being intermittent in part and occasionally showing diffraction contrast.

In thick crystals of the compound dislocation loops were once again formed. Unlike those in lead iodide, they were generally considerably larger and arranged in linear groupings, although some were of the usual size and randomly distributed. The large loops either took the form of separate loops along straight lines, with undamaged (to electron microscope observation) regions of crystal between these loops, or elongated loop configurations in the shape of double lines of dark contrast extending for distances of the order of 1 μ. These loop formations were completely random in direction in cases where the irradiation was by the normal method of placing the specimens in contact with the uranium foil (thus producing uncollimated fission fragments (Figure 66)). Where the bombardment was carried out using collimated fission fragment beams the lines of loops were parallel to one another, to within the degree of collimation (Figure 67). The density of the lines of loops was approximately the same as that of tracks in thin crystals, and was consistent with the radiation dose.

It might be thought that the alinement of the loops could be in a crystallographic direction, but this possibility can be eliminated for two reasons. Firstly, it is evident from Figure 67 that in the collimated fragment case the

---

Figure 63. Electron micrographs and diffraction patterns from a fission fragment irradiated crystal containing dislocation loops. In (a) a large number of reflections are operating simultaneously, while in (b) only the (0$\bar{3}$30) planes diffract strongly. The loop contrast disappears.

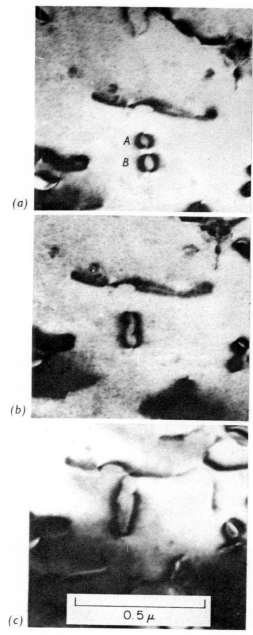

Figure 64. Electron micrographs showing the movement and coalescence of loops A and B in an area of fission fragment bombarded lead iodide during heating in the electron beam.

lines of loops are not exactly parallel, but have approximately the same degree of collimation as the incident particles. Secondly, in a comparison of micrographs of different crystals from the same support grid bombarded by collimated fragments, and of the appropriate diffraction patterns, it can be shown

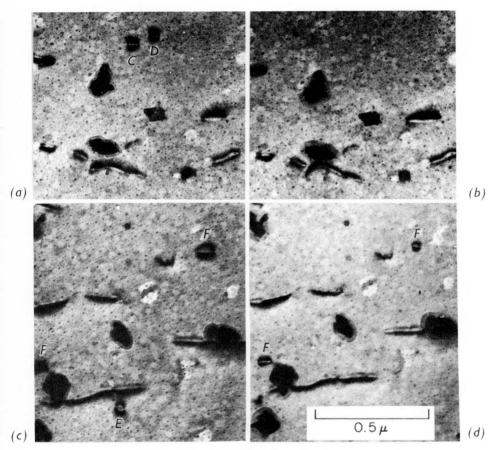

Figure 65. The sequences (a) and (b), (c) and (d) each consist of two consecutive micrographs of an area of irradiated crystal of lead iodide during electron beam heating. Loops C and D annihilate each other, loop E disappears, and the two loops F decrease slightly in size.

that the loop configurations are in the same direction in all cases, while the diffraction patterns show that the crystal orientations differ. There is, of course, an image rotation involved, but since this has the same value in all cases it is unnecessary to take it into account in a comparison of micrographs at the same magnification. We therefore conclude that these linear arrays of loops

result from another mechanism of track formation in thick crystals. Somewhat similar effects have been reported for irradiated crystals of platinum
(Ruedl, Delavignette, and Amelinckx, 1961).

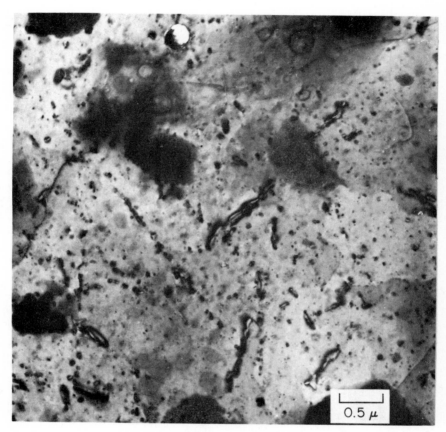

Figure 66. Electron micrograph showing linear arrangements of large dislocation loops in
thick crystals of cadmium iodide which have been irradiated by uncollimated fission fragments. The lines are random in direction.

Discussion

The heavy nuclei resulting from fission in a uranium foil we have generally
assumed (Chapter 1) to have a spectrum of energies which is binary and asymmetric (Glasstone and Edlund, 1952) with maximum numbers of fragments
having atomic numbers in the region of $Z = 39$ and $Z = 55$. These *median-light*
and *median-heavy* particles possess energies of 95 and 67 MeV respectively, and
have velocities in the region of 10^9 cm/sec. It has been demonstrated (Redmond,

Klingensmith, and Anno, 1962), however, that this energy spectrum is modified by changes in the thickness of the uranium foil, since the energy of a typical fragment is attenuated in travelling from its point of formation to the surface

Figure 67. An area equivalent to that of Figure 66, with the difference that the irradiation has been with collimated fission fragments. The linear arrays of dislocation loops are here approximately parallel, to within the degree of collimation.

of the foil. As the thickness of the foil approaches the range of the fragments in uranium (less than 10 μ), the double energy peak becomes single and shifts towards lower energies. Thus, for a foil of 3 μ thickness the energy peak is situated at about 30 MeV, and for the foils used in the present experiments, where the mean thickness was in excess of 20 μ, the majority of fission fragments had an energy of considerably less than 30 MeV.

Nevertheless, the thermal mechanism serves once again for a qualitative explanation of the type of damage which is visible in thin crystals of lead iodide, though other factors are also influential. In particular, the well-defined changes which occur in fission fragment tracks in crystals of lead iodide as the *thickness* is increased suggest that this is another very important parameter. The white tracks in very thin crystals are probably caused mainly by the thermal mechanism, where material in the region of the fragment path is raised to a high 'temperature' owing to the energy communicated to the lattice by excited electrons. If we consider that the observed track width is between 100 and 150 Å and assume that the track is approximately cylindrical in shape, then, since the crystals in question are around 50 Å in thickness, the region which is raised to the high temperature by a fragment travelling in a crystal almost parallel to the surface will extend to both top and bottom surfaces of the crystal. Consequently, the material inside the high temperature region is 'evaporated' through the surface, leaving an approximately linear hole in the crystal. Complications to this basic mechanism may arise from effects linked with the proximity of the surface. For example, the energy given to the lattice by excited electrons may be increased by multiple internal reflections of these electrons at the surface of the crystal (Noggle and Stiegler, 1960): thus the efficiency of track formation will be raised with decrease of crystal thickness to a greater extent than would be expected on the simple 'evaporation' model. The thermal mechanism of track formation in thin crystals is supported by a 'furrowing' effect, where the fragment enters and leaves the crystal. This takes the form of a region of reduced contrast at the end of a track, and is attributed to the fact that when the fragment is close to the crystal surface the heat generated can escape much more easily, and only the surface layers are removed. Therefore the visible image takes the form of a 'groove' which develops into a hole as the fragment penetrates the crystal.

As the crystal thickness increases above 150 Å, the material in the region of the track has greater difficulty in escaping from the surface, since these layers are now not greatly affected by the 'cooling' of excited electrons resulting from the passage of the fragment in the centre of the crystal. A less spectacular diffusion of heat to the surface then occurs, and some recrystallization of the central region and the creation of point defects ensues. Tracks in crystals of this order of thickness will therefore be made visible through absorption or diffraction contrast, or a combination of both, at the disordered core.

In many cases the types of mechanism described above may combine, resulting in the appearance of intermittent tracks which consist of elongated holes separated by regions of apparently undamaged crystal. The crystal thickness in these cases is in the region of or slightly greater than the observed track width, and their intermittency is probably due to effects such as variation of crystal thickness, surface irregularities, or variations in the amount of energy deposited per unit volume by the fission fragment.

It is possible that the 'spikes' observed on many white tracks (Figure 57) are due to a combination of thermal and displacement mechanisms, where secondary displacement cascades cause the high temperature region to extend over a much wider area. The features visible in tracks such as in Figure 60 may be due to a similar displacement mechanism. Experimental evidence supports the suggestion that most of the fragments from thick uranium foils have an energy (< 30 MeV) which is quite low in comparison with median-light and median-heavy particles. Also, the predominance of energy losses from nuclear collision and displacement is likely to begin at higher energies (~ 12 MeV) than might be expected from a strict application of classical ideas (Chapter 2). It is therefore suggested that the subsidiary branches clearly visible in Figure 60, and which are certainly associated with the main part of the track, may represent damage produced by primary knock-ons in a very thin crystal.

In thicker crystals, where it is difficult for material to escape to the crystal surface, white tracks due to the thermal mechanism do not form. At the same time the interstitials and vacancies formed along the tunnel created by the passage of the fragment are highly mobile, and the damage is not localized to the fragment path, so regions of strain in the core, which might give rise to black tracks through the operation of diffraction contrast, are not formed. Instead, these point defects move for relatively large distances through the crystal, tending to coagulate as dislocation loops randomly distributed throughout the specimen. Diffusion and clustering of interstitials can take place with ease between different (0001) sandwiches of lead and iodine atomic planes, and also, it is suggested, between members of the molecular $\{10\bar{1}1\}$ family of inclined pyramidal planes. Under both these mechanisms the formation of dislocation loops in the observed plane is possible. Further evidence in support of these suggestions is presented in the following chapter.

From the appearance of the tracks in cadmium iodide it is evident that the thermal mechanism of track formation in thin crystals operates similarly in this case, material escaping through the surface of the crystal in the wake of each highly ionizing energetic particle. The close similarity between the iodides of cadmium and lead presents some problem, however, in obtaining an explanation for the difference between the visible damage in *thicker* crystals of the two materials. Why are the dislocation loops in the lead compound apparently distributed throughout the irradiated crystal with no dependence on the direction of the incident particle, whereas in cadmium iodide they appear in linear groupings along the path of the energetic fragment? The solution to the problem may lie in the mobility of the large number of point defects created by excitation and ionization along the tunnel formed by the passage of the fragment. It is thought that the coagulation of these and their interaction with the lattice leads to the formation of the dislocation loops: the defect mobility is therefore an important factor in deciding how far from the defects the loops are formed. If a difference in defect mobility were a prime cause of the difference between the visible signs of damage in the two compounds, it would appear that the defects are more mobile in lead iodide, travelling a considerable distance from their point of formation before coagulating into loops, whereas in cadmium iodide their mobility is smaller, and coagulation occurs close to the point of formation. The slight difference in structure between the cadmium and lead iodide lattices, resulting from the different stacking of the molecular sandwiches in the two cases, reduces the space between members of the $\{10\bar{1}1\}$ planes in the cadmium compound. The $\langle 10\bar{1}1 \rangle$ direction is considered to be important for interstitial migration in this lattice, and minor differences in the structure of this 'channel' may well lead to a major reduction in point defect mobility in cadmium iodide.

Conclusion

An electron microscope study of the damage visible in crystals of lead iodide which have been irradiated by fission fragments has revealed a sensitive dependence of the type of damage on crystal thickness. The efficiency of track formation depends on the proximity of the crystal surface, as well as on the rate of energy loss of the incident particle. The change in appearance and eventual disappearance of tracks as the crystal thickness increases may be qualitatively explained on the basis of a thermal mechanism. The damage configuration in thick crystals indicates that the point defects initiated by the fission fragments are highly mobile, and the orientation of the dislocation loops and the lattice structure suggest some possible crystallographic directions in which these point defects may move with greatest ease.

In a similar experimental study of damage in cadmium iodide an additional mechanism of track formation has come to light, in which the path of the particle is revealed by the alinement of dislocation loops in the direction of travel. This seems to indicate a higher degree of localization of damage in cadmium iodide, possibly associated with a smaller point defect mobility.

References

BROSTRØM, K. J., BØGGILD, J. K., and LAURITSEN, T. (1940), *Phys. Rev.* **58**, 651.

CHADDERTON, L. T. (1962a), *J. Appl. Phys.* **33**, 3021.

CHADDERTON, L. T. (1962b), *Nature*, Lond. **195**, 987.

CHADDERTON, L. T., and TORRENS, I. MCC. (1966), *Proc. Roy. Soc. A* **294**, 80.

DAWOOD, R. I., and FORTY, A. J. (1962), *Phil. Mag.* **7**, 1633.

FORTY, A. J. (1960), *Phil. Mag.* **5**, 787.

FORTY, A. J. (1961a) *Phil. Mag.* **6**, 587.

FORTY, A. J. (1961b), *Phil. Mag.* **6**, 895.

GLASSTONE, S., and EDLUND, M. C. (1952), *The Elements of Nuclear Reactor Theory*, Van Nostrand (New York).

HIBI, T., and YADA, K. (1960), *J. Electron Microsc. Japan* **9**, 101.

HIRSCH, P. B., HOWIE, A., and WHELAN, M.J. (1960), *Proc. Roy. Soc. A* **263**, 217.

HORNE, R. W., OTTEWILL, R. H., and MCMORRIS, M. (1961), *Proc. AIME Tech. Conf. Direct Observation of Defects in Crystals*, St. Louis, p. 239, John Wiley (New York).

NOGGLE, T. S., and STIEGLER, J. O. (1960), *J. Appl. Phys.* **31**, 2199.

PRICE, P. B. (1961), *Phil. Mag.* **6**, 449.

REDMOND, R. F., KLINGENSMITH, R. W., and ANNO, J. N. (1962), *J. Appl. Phys.* **33**, 3383.

RUEDL, E., DELAVIGNETTE, P., and AMELINCKX, S. (1961), *J. Appl. Phys.* **32**, 2492.

SILCOX, J., and WHELAN, M. J. (1960), *Phil. Mag.* **5**, 1.

TORRENS, I. MCC. (1964), *Proc. 3rd Europ. Reg. Conf. Electr. Microsc. A*, Prague, p. 193.

CHAPTER TEN

Atomic Displacement in the Lead Iodide Lattice

Our consideration of fission fragment damage to various crystal types has so far been confined largely to microscopic observations of effects produced by single particles. We now turn our attention to a close examination of those fundamental processes which give rise to the features observed.

We shall see later (Chapter 12) how the type of crystal binding and the electronic mobility exert considerable influence on the permanent damage configuration of the crystal following electron excitation by the passage of an energetic fragment. Important as this electronic energy attenuation indisputably is, we clearly may not neglect that portion of the loss which may be attributed to direct collisions with the atoms or nuclei of the crystal (equation 2.21). This may arise through close collisions of the fission fragment with crystal nuclei. However, the probability of such events is small; we should expect only one such collision in more than 10^6 Å of the fragment path in the crystal. More important from the point of view of collision processes are the secondary effects of the passage of the fragment. A large number of displaced atoms are left in its wake, originating primarily in regions of high lattice excitation. These knocked-on atoms may have energies from a few electron volts up to thousands and their subsequent histories are of paramount interest since they play a significant role in determining the final damage configuration in the crystal.

In the previous chapter, mention was made of some electron microscope observations of damage to lead iodide which indicated considerable mobility of the resulting displaced atoms in the lattice. We propose here to extend the investigation of damage in the lattice of this particular crystal to encompass atomic displacement and its influence on the final damage pattern. In doing so we shall introduce a technique which has been employed by a number of investigators – computer simulation of a regular crystal lattice array (Chadderton and Torrens, 1966).

What may we expect to happen in the lead iodide lattice when an atom receives energy sufficient to remove it from its equilibrium position? We could, in principle, calculate analytically the subsequent trajectory of a knocked-on atom in the lattice, its interaction with its neighbours, and the spread of the disturbance in the crystal. In practice, however, this project is extremely formidable, requiring the simultaneous integration of a great many differential equations of motion. Fortunately we have available a very useful tool for the purpose in the high-speed digital computer. By employing iterative processes it is possible to follow complicated damage events as they develop from a prescribed initial disturbance. We shall see how such computer simulations demonstrate the influence of directional effects (focusing and channelling) in the hexagonal lead iodide lattice.

Method of Computation of Primary Displacement Events

The simulations were of a similar nature to those performed at Brookhaven National Laboratory (Gibson, Goland, Milgram and Vineyard, 1960) on metals, enlisting the aid of a high-speed digital computer to follow a many-body collision process through consecutive small increments of time. Because of computer time and space limitations our calculation for lead iodide was limited to the motion of the atoms in one plane of the crystal. The initial positions of the atoms in the array were fed into the machine, and one atom was given a kinetic energy. The computer then employed the classical equations of motion to find the positions and velocities of the atoms at the end of consecutive time steps. A central difference procedure was employed to integrate the differential equations of motion.

From Newton's second law we have:

$$[v_i(t) = (1/M) F_i[x_1(t), x_2(t), \ldots, x_n(t)] \tag{10.1}$$

where M is the mass of the atom, $x_i(t)$ and $v_i(t)$ are the ith atomic coordinates and velocities respectively at time t, and F_i is the force in the ith degree of freedom. With two-dimensional Cartesian coordinates n is twice the number of atoms in the array considered. When x_i refers to a boundary atom, F_i depends also on the velocity v_i due to viscous damping.

Approximating by finite differences, we have:

$$v_i(t + \tfrac{1}{2}\Delta t) = v_i(t - \tfrac{1}{2}\Delta t) + (\Delta t/M) F_i \tag{10.2}$$

$$x_i(t + \Delta t) = x_i(t) + \Delta t v_i(t + \tfrac{1}{2}\Delta t) \tag{10.3}$$

Thus, by knowing all velocities $v_i(t - \tfrac{1}{2}\Delta t)$ and positions $x_i(t)$ at time t and from calculating the interatomic forces F_i, the new positions and velocities after a time Δt can be computed. These values may then be taken as initial values for the next time-step, and so on.

The time interval Δt which was used in the computation depended on the velocity of an atom, and was small in the early stages of the calculation (typically 0·1 units), increasing as the velocities of the atoms decreased. Checks on energy conservation were incorporated into the programme.

In order to facilitate analysis of the results, units were chosen as follows:

length $L = 4{\cdot}54 \times 10^{-8}$ cm
time $T = 1{\cdot}488 \times 10^{-14}$ sec
mass $M = 1{\cdot}720 \times 10^{-25}$ g
energy $E = 1{\cdot}602 \times 10^{-12}$ erg $= 1$ eV
velocity $v = 3{\cdot}052 \times 10^{6}$ cm/sec

The unit of length was the basal plane interatomic spacing of lead iodide, the unit of energy was 1 eV, and the unit of time was such that a lead atom with 1 keV of energy had unit velocity. The units of mass and velocity followed from these definitions. It should be noted that the lead atom has a mass of 2000 in this system of units.

Initially, some runs were made on the basal (0001) plane of lead atoms. An array of 100 atoms in this plane was chosen, and the coordinates and velocities of these were read into the machine, a two-dimensional Cartesian system being used.

To investigate the motion of knock-ons in a plane containing both lead and iodine atoms, the $\{1\bar{2}10\}$ vertical plane was chosen, and a larger array, containing 420 atoms in the form of 14 molecular sandwiches each with 30 atoms, was used.

The Interatomic Potential

A potential of the Born-Mayer type $[V(r) = A\exp(-Br)]$ was considered to be the most accurate at distances corresponding to impact parameters for collisions involving energy transfers of up to 1 keV. Since lead iodide possesses a hexagonal lattice, and contains more than one type of atom, it is not possible to compute the Born-Mayer constants directly from elastic constant data. The constant involved in the exponential (B) was therefore calculated by extending the Brinkman treatment (Brinkman, 1960):

$$B = [\sqrt{(Z_1 \; Z_2)}]^{1/3}/(Ca_0) \qquad (10.4)$$

$$a_0 = 0{\cdot}5282 \times 10^{-8} \qquad (10.5)$$

where Z_1 and Z_2 were the atomic numbers of the two colliding atoms, and C was a constant with the dimension of reciprocal centimetres, chosen to be unity. The constant A was given the value of 10^8. For a compound with a structure as complicated as that of lead iodide, it is very difficult to choose a potential which is certain to be a reasonable approximation to that which actually operates. With the potential chosen, however, the array was stable within the tolerance of the calculation, and, in the computations, stress was placed on obtaining a qualitative model for the behaviour of knocked-on atoms with energies which might result from primary fission fragment collisions, rather than on accurate quantitative results.

Since the Born-Mayer potential $V(r)$ is repulsive for all r, an attractive term was required to reproduce the crystal binding forces. This was obtained using the Coulomb attraction of ions of opposite charge. A Coulomb potential was used, which had the value of 3·5 eV at the equilibrium position of the lead and iodine atoms (i.e. 3·13 Å) and which was derived from the binding energy of lead iodide, dependent not only on the ionic attraction, but also on the high degree of covalent binding of the compound. The actual potential was a

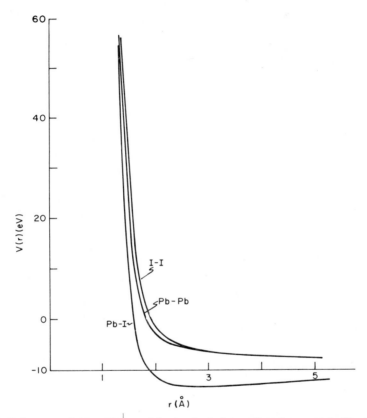

Figure 68. Interatomic potentials used for two-body interactions in a crystal of lead iodide. In the case of Pb–I interaction, the Born-Mayer repulsive potential has an attractive tail at separations in the region of the Pb–I spacing, to simulate the molecular binding.

numerical potential which behaved as the sum or difference of the Coulomb and Born-Mayer types, according to whether the interacting ions were of the same or opposite charge. The curves for each interaction are shown in Figure 68. Since these are empirical, and contain contributions from covalent binding, they should not be compared with conventionally accepted values of ionic radii, which are based on pure ionic binding.

The Programme

Given a set of coordinates and velocities for a specified number of atoms in a crystallographic array, the basic problem was to calculate the force on any one atom due to its interactions with all the remaining atoms of the set. This can be accomplished, but for a 420 atom array it would involve the calculation of more than 104,000 forces for only one time-step, and the computer time required for such an operation would be prohibitive if one wished to continue the calculation for more than a few time-steps. Several measures were taken to cut down the computer time required per time-step of the calculation, subject to the condition that none of these significantly altered the course of the damage event.

The atoms in the set, numbered from 1 to 420, were each ascribed a function $\mathscr{F}$ which was zero if the atom was considered 'at rest', and unity if it was moving. Thermal and zero-point motions were neglected and the set was assumed to be at a temperature of $0°K$. An atom was considered to be at rest at the beginning of any time-step if the force on it due to all of the moving atoms during the preceding time-step was such that its displacement from its position at the beginning of that step would be less than 10^{-4} unit of length. At the beginning of the damage event one atom was set in motion in a particular direction with the required velocity by giving it the appropriate values of x and y velocity. Simultaneously, its function $\mathscr{F}$ was set equal to one, with that of all the remaining atoms zero. The computer was asked to consider all those atoms for which $\mathscr{F}$ was equal to 1, in relation to their interactions with all other atoms. It selected each of these in numerical order and checked the difference between its x and y spatial coordinates and those of the remaining atoms in the set. If these differences were greater than 2 lattice units, or in other words if one of the remaining atoms A_j was outside a square of side 4 units centred on the moving atom A_i, the interaction between A_i and A_j was neglected. If A_j was inside the square the distance between A_i and A_j was computed, and if this was greater than 1·5 lattice units of length, the interaction was again neglected. The justification for these operations was the knowledge that the interatomic potential at distances greater than 5 or 6 Å is sufficiently small to allow a cut-off in this region, without any significant decrease in accuracy.

The force acting on A_i due to the atoms in its neighbourhood (less than 1·5 units distant) was then computed directly from the gradient of the interatomic potential:

$$F_r^{(i)} = -dV/dr \tag{10.6}$$

As was previously stated, the computer stored this potential in numerical rather than analytical form, and $V(r)$ and $F_r^{(i)}$ for any r could be found by a standard interpolation library routine. Once calculated, $F_r^{(i)}$ was resolved into its x and y components, and the total $F_x^{(i)}$ and $F_y^{(i)}$ acting on A_i owing to its interaction with all neighbouring atoms A_j was found by summation. These, then, were the forces to be used in the equations 10.2 and 10.3. If the force between A_i and any A_j was found to be so large that during the following time-step A_j could no longer be considered at rest, that is if the force between A_i and A_j satisfied the condition:

$$F_{x,y}^{(ij)} (\Delta t/M) > 10^{-3} \qquad (10.7)$$

then the $\mathscr{F}$ function of A_j was changed to 1, and it was classified as a moving atom for the next time-step. Also, if both $F_x^{(i)}$ and $F_y^{(i)}$ were found to be less than 10^{-4}, and if the velocity of A_i was negligibly small for the purposes of the calculation, its $\mathscr{F}$ function was made zero. To avoid confusion during the current time-step, however, the value of $\mathscr{F}$ had to be held at its current value until the beginning of the next step, and then set to its new value.

Having found $F_x^{(i)}$ and $F_y^{(i)}$ and checked whether they were greater than 10^{-4} units by summation of all the $F^{(ij)}$ forces, equations 10.2 and 10.3 were used to generate new position and velocity coordinates for A_i. These values also were stored until the end of the time-step before being substituted for the old set. The computer, on completing this operation for atom A_i, investigated whether other atoms in the array were in a state of motion or stationary, by changing the value of i; if $\mathscr{F}$ was non-zero for any atom, the sequence of operations was repeated for that atom, and its interactions with nearby atoms were calculated. When this process had been carried through for all of the atoms in the set, the new positions, velocities, and $\mathscr{F}$ values were substituted for the old ones, and they became the initial conditions for the next time-step.

The forces on the edges of the array due to the interior atoms required to be balanced by external forces which simulated the effect of an infinite plane of atoms extending beyond the edges. These forces, like those in the centre, depended on the displacement of the atom from its lattice position, and on its velocity, thus allowing energy loss from the array. The position-dependent force took the form of a Hooke's law restoring force:

$$F_r = -kr \qquad (10.8)$$

and was allowed to act only when r became greater than a certain value. This value was determined by adjustment of the value of k until the behaviour of an atom projected directly from its lattice position in one of the edges was closely similar to that of an equivalent atom in the centre of the array. The same procedure was followed to obtain the constant in the velocity-dependent force:

$$F_v = -k'v \qquad (10.9)$$

Since the atoms at the edges of the set were arranged in closely situated but offset planes of iodine and lead the restoring forces should not strictly be the same for all these atoms, and they should also have a directional dependence. The picture here becomes extremely complicated, and in view of this it was considered sufficient to stabilize the array by approximately correct forces

11

directed inward. The inaccuracy of this did not significantly affect atoms more than one lattice spacing from the edge, and damage events were also kept to the central region of the set as far as was possible. During the actual runs, the surface atoms were only very rarely displaced from their lattice positions, usually with low energies.

At the end of each time-step, an energy check was performed to ensure that the balance between potential and kinetic energy was maintained to within reasonable limits. In an attempt to avoid the propagation of machine errors, a check on the order of the interatomic forces was incorporated; if any atom received a displacement which was unreasonably large, the computer re-calculated the time-step, and, if the result was repeated, the calculation was stopped.

At specified intervals of a few time-steps, the positions and velocities of the moving atoms and the initial and final kinetic and potential energies were printed out. On some runs the positions at each step were displayed on a curve-plotter coupled to the computer, but although this saved effort after the run it was time-consuming during the computation.

Choice of Crystal Plane

A number of runs were made in the (0001) plane of lead atoms, and the energy of the initial knock-on was varied between 20 and 1000 eV. The potential employed was in this case a simple Born-Mayer type without any Coulomb term, since the atoms were all of the same kind. An atom moving in this plane experiences forces due to atoms in the iodine planes situated at a distance of 1·72 Å above and below, and in its equilibrium position these are stronger than the forces on it due to neighbouring lead atoms. These forces are, in fact, the molecular binding forces. Thus to obtain a more accurate picture of the motion of displaced atoms in the (0001) plane it is necessary to calculate these inter-planar forces. Because of computer space and time limitations this has not so far been possible. It is thought that atoms moving in the (0001) plane will have a high probability of being deflected out of the plane by atoms of close parallel planes into channels either between the sandwiches or at an angle to the basal plane. Confirmation of this has been obtained from calculations of events in a vertical plane.

Several possibilities presented themselves as candidates for simulation studies in a vertical plane. Of these the $\{1\bar{2}10\}$ set was chosen, mainly because of its molecular constitution and the constant separation between neighbouring planes of the set. The molecules were in equilibrium with the line joining the component nuclei within each molecule lying at an angle of approximately 33° to its projection on the basal plane. The separation between the lead atoms of adjacent molecules was 6·86 Å in a vertical direction, and 7·86 Å horizontally. As a consequence of the molecular binding, it was necessary to use a potential incorporating an attractive term. The potential shown in Figure 68 was chosen.

During calculations with this plane, the computer required to know whether the interacting atoms were lead or iodine, and this information was stored again in the form of a quantity which was zero for iodine atoms and unity for lead atoms: this quantity did not, of course, alter during the computation.

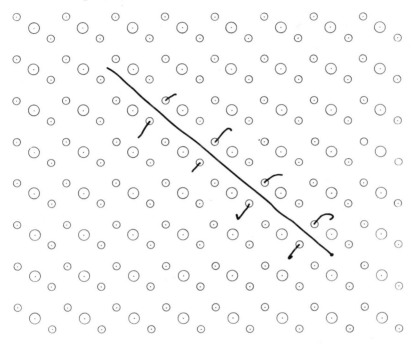

Figure 69. Simulation of a damage event in the (1$\bar{2}$10) plane of lead iodide, in which an interstitial lead atom is given an energy of 50 eV and set moving directly along the axis of a [10$\bar{1}$1] channel. This diagram demonstrates the perturbation of the atoms in the walls of the channel, showing the I–I potential barrier which is mainly responsible for attenuating the energy of a perfect channelon. The set of atoms shown in this and further diagrams is only a section of the 420 atom array used in the computation. ◯, Lead atom; ○ iodine atom.

Some of the events studied in this simulation were introduced for the purpose of investigating channelling possibilities in $\langle 10\bar{1}1 \rangle$ and $\langle 10\bar{1}0 \rangle$ channels. The former are channels bounded by molecules of the corrugated $\{10\bar{1}1\}$ planes and their axes lie at an angle of about 41° with the projection on the basal plane. The $\langle 10\bar{1}0 \rangle$ channels are parallel to the basal plane, being essentially the space between the I–Pb–I sandwiches. During these events, interest was mainly centred around the channelling atom or channelon (Chadderton, 1965), which was initially an interstitial introduced deliberately at a convenient point in the lattice and given a kinetic energy of up to 1000 eV. Consequently, to save machine time, it was arranged that only data concerning the channelling atom should be printed. Thus in some of the diagrams showing atomic trajectories, while only one atom is shown to be moving, it must be remembered that at a number of collisions this atom influences atoms in the walls of the channel, and although no data were printed for these atoms, they were nevertheless permitted to move during the calculation. A specimen event showing how a low energy channelon affects the atoms in the walls of the channel is depicted in Figure 69.

Channelling in the $\langle 1011 \rangle$ Direction

Events were run at energies between 80 and 1000 eV (Figures 70(a) and (b)) with the channelon trajectory making an angle of 0 to 20° with the axis of the channel. The subsequent behaviour of interstitial atoms set moving at various angles to the channel direction clearly indicated that, in lead iodide, channelling is a much higher energy phenomenon than focusing. Channelons with an energy of 80 eV required to be projected along or within a few degrees of the channel axis if they were to survive as such for more than 20 to 30 Å (Figure 70(a)); the reason for this was that when they suffered collisions with atoms in the walls of the channel they lost a high proportion of their energy. Interstitials of this energy projected along or at 2° to the axis channelled very efficiently, since they effectively did not suffer any Born-Mayer type collisions with the sides of the channel, and they could theoretically travel for a long distance down the channel, being restricted to the centre by the longer-range weak Coulomb forces. They did tend to lose energy gradually, however, mainly as a result of passing through the 'neck' of iodine atoms, which two atoms were only 4·32 Å apart and formed the narrowest section of the channel. These relaxed slightly as a result of the passage of the channelon (Figure 69). The other two pairs of atoms, both Pb–I pairs, caused some variation in the energy of the axial channelon, but this subsidiary variation was small compared with that due to the I–I pair.

A plot of kinetic energy against distance travelled in the channel (in units of the basal plane spacing, $r_0 = 4·54$ Å) for a lead interstitial of initial energy 80 eV shows how sensitive a channelon of such a low energy is to all three barriers (Figure 71(a)). From the graph it can be seen that the barrier, while impeding the progress of an approaching channelon, accelerates that same atom

---

Figure 70. (a) Trajectories of 80 eV channelons projected at various angles to the channel axis. For all but very small angles, dechannelling occurs rapidly. In this and in Figures 70(b) and 72, the trajectory of the channelon only is shown; the lattice atoms are however allowed to relax during the computation. (b) Some 1000 eV lead atom channelling events. A considerable variation of initial direction of motion is permitted for energies of this order. Even 20° events have a range ~ 50 Å. For small angles the forward motion has a damped transverse oscillation superimposed, which in a perfect lattice should eventually tend to zero. The dotted line affords a comparison with the equivalent 10° event run with the lead atom replaced by an iodine atom. This, as can be seen from the figure, has a shorter channelling range, since the amplitude of its transverse oscillation gradually increases.

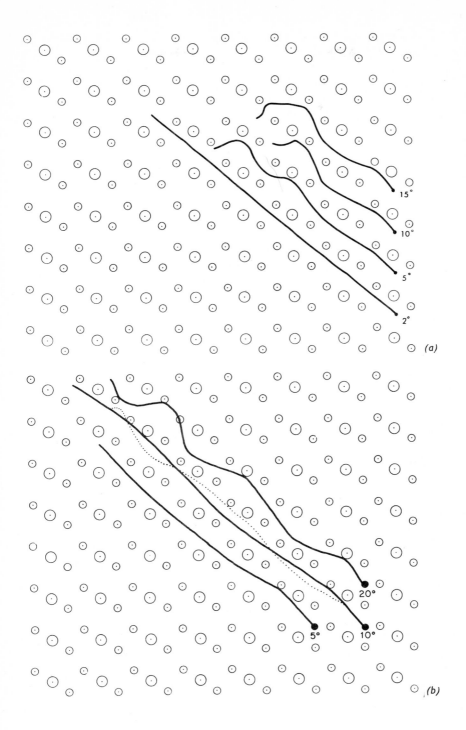

(a)

(b)

after it has passed through; so that the overall effect of the barrier is to decrease the energy of the atom by about 1 %, although temporarily the moving atom loses about 5 % of its energy before receiving a compensating impulse. The total energy loss on travelling a distance of $10r_0$, or about 45 Å, is approximately 1·5 eV, which would give the channelon a theoretical range of up to 2500 Å, barring accidents such as the presence of an impurity or an imperfection of the lattice in its path.

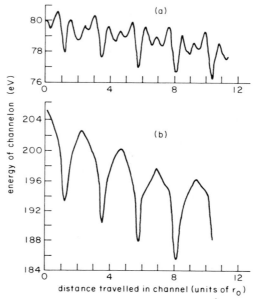

Figure 71. Comparison of the variation of energy of a lead channelon with distance travelled along the channel (in units of the basal plane spacing $r_0 = 4\cdot54$ Å) for atoms of initial energy: (a) 80 eV and (b) 205 eV. The sensitivity of the 205 eV channelon to the potential barriers due to the atoms in the walls is much less than that of the 80 eV atom, and two of the barriers of the latter case are almost completely smoothed out in the higher energy event. A 1000 eV channelon shows no detectable periodic variation of energy with distance.

It must be remembered, however, that we have been considering the idealized case of an atom travelling directly along or at a very low angle to the channel direction, and that the majority of 80 eV channelons have a practical range of about 25 Å. As the initial energy of the channelon is increased, the channelling efficiency might also be expected to increase rapidly, since the channelon will be less sensitive to collisions with atoms of the walls. This point may be illustrated by the graph of channelon energy variation with distance for a 205 eV channelling atom (Figure 71(b)). The subsidiary energy peaks due to the Pb–I barriers have almost completely disappeared, and only the I–I barrier influences the channelon energy. The overall energy loss of the channelon on passing through this barrier is again about 1 %, similar to the

80 eV case. A similar run with a 1000 eV channelling interstitial revealed no sensitivity whatever to any of the atomic barriers; any substantial energy loss of such a perfect channelon must necessarily be due to interaction with phonons and imperfections in the lattice. However, an atom with this energy was of substantial interest since it appeared to channel quite efficiently even if set in motion at angles of up to 20° with the channel axis, one such atom travelling an estimated distance of some 50 Å before being knocked out of the channel. The paths of atoms which did not move a great distance from the axis took the form of an oscillation of long period and small amplitude about the axis. For larger angles, where the interaction with the walls was greater, the oscillation was more irregular, tending towards a shorter period and a larger amplitude, until the atom was finally removed from the channel by a close collision with an atom in the wall.

Channelling in the ⟨10$\bar{1}$0⟩ Direction

The space between the walls of these horizontal channels is less than that of the oblique ones, being 3·43 Å. The forces on a channelon due to atoms on either

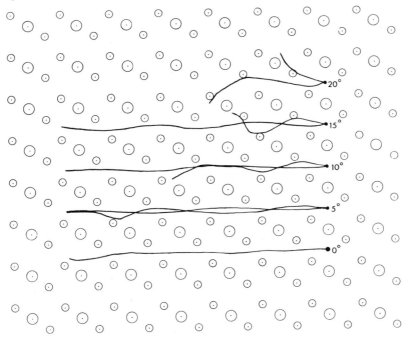

Figure 72. Channelling in the [10$\bar{1}$0] direction, between the basal plane sandwiches. The asymmetry of the channel can be seen from a comparison of events with the same angle of projection above and below the channel axis, and from the transverse oscillation of the axis channelling (0°) event.

side are not balanced, so that the trajectory of an atom initially moving directly along the axis is not a straight line but rather a progressive transverse oscillation,† whose period and amplitude remains constant. If the channelon is initially moving off-axis, the oscillation tends to decrease in amplitude until it resembles that for an atom travelling along the axis. Some trajectories of 1000 eV [10$\bar{1}$0] channelling lead atoms are shown in Figure 72. The energy variation of a channelon in this direction was somewhat erratic, as a result of its transverse oscillatory motion and irregular collisions, but there was an overall gradual loss of energy which amounted to approximately 60 eV in a distance of 50 Å, rather more than that of the equivalent [10$\bar{1}$1] channelon. The same oscillatory movement, on the other hand, sometimes caused quite a high channelling efficiency for atoms projected at angles of 10° and greater to [10$\bar{1}$1].

Focusing in the {1$\bar{2}$10} Plane

The focusing phenomenon was expected to be less prevalent in an open lattice of this type than in a tightly packed lattice, since defocusing could occur without much restriction. Events were run which involved projecting an atom of the plane with energy 80 eV at angles varying from 0 to 360° with [10$\bar{1}$0]. One such event is shown in Figure 73, where the angle of projection is 20°. During this, as in most of the other runs, some subsidiary focusing of energy took place in the vertical [0001] directions, but this was generally short-lived, owing to the asymmetry of the adjacent lines of atoms. It is interesting to note that an iodine atom displaced in a secondary collision becomes (at least temporarily) channelled in the [10$\bar{1}$1] direction. Because of its low energy, however, it will probably be stopped before travelling very far. It is difficult to assess the probable final damage stage of this region of the plane, since the relaxation period in the lattice is relatively long, though a number of replacements are probable, and there are many possible stable interstitial configurations.

The only possible direction in the plane which provides for strongly assisted focusing is the [$\bar{1}$01 1] direction, along a line of lead atoms. This direction is approximately perpendicular to (10$\bar{1}$1) planes of *alternating atomic constitution*, as distinct from those containing both lead and iodine atoms. A lead atom travelling in this direction passes through an assisting lens of two iodine atoms which aids the focusing. However, the distance it must travel before colliding with its successor in the chain is 10·43 Å, and consequently it does not require to be more than a few degrees off-axis in order to defocus

---

† This oscillation perpendicular to the basal plane illustrates the asymmetry of the forces acting on a horizontal plane due to planes above and below – the principal reason why simulations of the (0001) plane are inaccurate and unrealistic.

rapidly, owing to over-correction by the I–I lens. This is another consequence of the high lattice spacing.

Figure 74 illustrates graphically some of the above points, by means of the super-position of atomic trajectories following the projection of 80 eV primary atoms at angles from 0 to 360°, at intervals of 20°. This we call a *correlation diagram*. The temporary focusing and rapid defocusing along vertical lines can be seen, together with the strong, but very selective, [$\bar{1}011$] focusing. For all runs other than the 40° one, focusing in this direction is restricted to one collision. The large ranges of the displaced atoms and the lack of regularity of the spread of the region of the disturbance should be compared with the correlation diagram for a more regular, tightly packed, cubic lattice (Figure 85, Chapter 11).

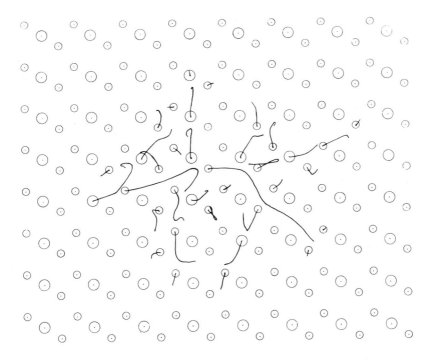

Figure 73. Progress of an event in which an 80 eV lead atom is set moving at an angle of 20° to the basal plane. Subsidiary focusing chains are initiated in the [0001] directions, but these are short-lived owing to the asymmetry of the adjacent lines of atoms. An iodine atom becomes at least temporarily channelled in a ⟨$10\bar{1}1$⟩ channel.

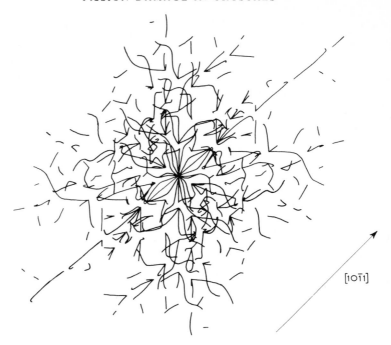

[1$\bar{0}$1̄]

Figure 74. The 80 eV energy correlation diagram for the (1$\bar{2}$10) plane, which superimposes trajectories of the atoms in events where the primary lead atom is projected with energy 80 eV at angles varying from 0° to 360°. Strongly assisted focusing occurs in the 40° event for atoms moving in the [$\bar{1}$011] direction, but this requires projection within a small angle of this direction, otherwise defocusing takes place immediately. The relatively long range of the secondary knock-ons is a consequence of the very open lattice of lead iodide.

Channelling of Lattice Atoms

The computed events have clearly shown that an atom with an energy of the order of 1000 eV, on finding itself in a favourable position and moving in a suitable direction in the lead iodide lattice, can travel large distances along $\langle 10\bar{1}1 \rangle$ channels. But does an atom knocked-on from its lattice position with this energy have a reasonable probability of being deflected into a channel? To obtain information on this, events were investigated in which a lead atom, initially at rest in a lattice position, was set in motion at directions making angles ranging from 0 to 360° with a vertical [0001] direction, and the subsequent trajectories of this atom were followed. The procedure was repeated for a lattice atom of iodine. The trajectories of these *primary knock-ons* only were then superimposed on a *primary energy correlation diagram*, in order to display the polar efficiency of the channelling process. For example, the primary energy correlation diagram for an 80 eV event, involving a lead atom, is obtained by removing all the background of secondary atomic orbits from Figure 74,

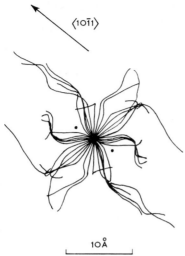

Figure 75. An 80 eV primary energy correlation diagram for a lead atom in lead iodide, which results from removing all the secondary atomic orbits from Figure 74. There is no significant variation in length of orbit with angle of projection at this low energy, although the $\langle 10\bar{1}1 \rangle$ channels are beginning to exert their influence.

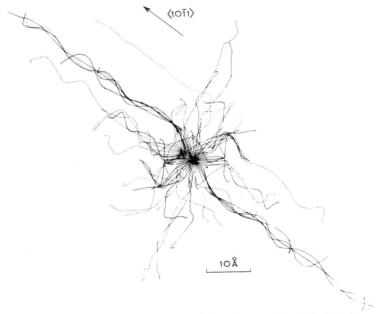

Figure 76. Superimposed primary energy correlation diagrams for 1000 eV lead (continuous lines) and iodine (dotted lines) atoms. These show a marked variation in length of orbits, with $\langle 10\bar{1}1 \rangle$ channelling occurring strongly in the case of the lead atom. The 1000 eV iodine atoms have a smaller range than equivalent lead channels, and tend to be deflected after several collisions into secondary $\langle 10\bar{1}1 \rangle$ channels several lattice units away from the primary event.

leaving the trajectories of Figure 75. We note that in this low energy case the orbits of the knocked-on lead atom do not vary significantly in length, although even for an 80 eV primary the $\langle 10\bar{1}1 \rangle$ channels are beginning to demonstrate their existence. If we now plot an equivalent diagram for a 1000 eV knock-on (Figure 76), it becomes evident that there is a very strong angular variation in length and shape of the orbits. In the lead atom case, channelling occurs in the $\langle 10\bar{1}1 \rangle$ directions with high probability within two angular bands of about 30°. The range of 1000 eV channelled iodine atoms is smaller than that of the equivalent lead channelons, and these lighter atoms show a tendency to be deflected, following several collisions, into secondary $\langle 10\bar{1}1 \rangle$ channels not immediately adjacent to the primary event. The difference in behaviour of the lead and iodine atoms projected under the same conditions is demonstrated by superimposing the primary energy correlation diagrams for the two cases, with the original knock-ons in their relative positions in the lattice (Figure 76).

Note on the Validity of the Two-Dimensional Approximation

It is thought to be reasonable in the case of the vertical planes to consider the effect of the atoms in one plane only in their interactions with a moving atom, since the neighbouring planes above and below are complementary. The forces on an atom moving in the plane do not tend to remove it, as they do in the case of the basal plane. A three-dimensional treatment would reveal some minor differences: for example, during $\langle 10\bar{1}1 \rangle$ channelling, the energy loss on passing through the I–I constriction would be increased, since in three dimensions this constriction consists of four atoms – two iodine atoms in the $\{10\bar{1}1\}$ plane and two other lead atoms above and below the plane, and equidistant from it. The I–I and Pb–Pb distances are approximately the same, the lead atom pair lying in the basal plane, separated by the lattice spacing of 4·54 Å. Hence the energy loss of a perfect channelon passing through this barrier is probably doubled. In passing through the two Pb–I barriers of this channel, the energy dissipation to atoms of neighbouring planes is greater than that to atoms of the $\{1\bar{2}10\}$ plane of the simulation, so the two-dimensional treatment probably overestimates the range. The oscillation about the channelling axis of an imperfect channelon, if extrapolated to three dimensions, becomes a somewhat irregular spiralling movement since the 'lenses' are rather astigmatic.

Thus, while quantitatively the neglect of the effect of adjacent planes above and below the simulation plane leads to a little inaccuracy, we nevertheless can achieve a very good qualitative picture of channelling and focusing in the lead iodide lattice, and it is not difficult to make reasonable semi-quantitative estimates of the range and behaviour of channelling atoms. This conclusion has been confirmed by subsequent computations in a simulated three-dimensional potassium chloride lattice to be described later, where we shall find that the atomic motions within a plane are very little altered by the presence of atoms in planes above and below. The performance of a three-dimensional study for lead iodide is presently prohibited by the demands which the complexity of the crystal structure would make on computer time.

Conclusion

Computer simulations in the vertical $\{1\bar{2}10\}$ plane of the lead iodide lattice have therefore shown that channelling of atoms might be an important source of energy loss from the region of a typical primary event, such as would occur in many types of fission damage experiment. The probability of channelling in the $\langle 10\bar{1}1\rangle$ and $\langle 10\bar{1}0\rangle$ direction is high over a considerable range of energy and angle of projection. In an experiment where crystals of lead iodide are bombarded by fission fragments a large number of primary knock-ons of this order of energy are initiated in the region of the path of a fragment, and many of these are suitable candidates for channelling. In the preceding chapter it was concluded that the majority of the dislocation loops observed in the electron microscope following fission fragment irradiation lay on $\{10\bar{1}1\}$ planes. The $[10\bar{1}1]$ channel investigated during the simulation of events in a $\{1\bar{2}10\}$ type plane is bounded in three dimensions by four corrugated lines of atoms in $\{10\bar{1}1\}$ planes. This channel, in three dimensions, becomes a tunnel in the lattice bounded by these four lines or, from another point of view, bounded by four intersecting planes which are all members of the $\{10\bar{1}1\}$ family. A possible mechanism, therefore, for the formation of the random array of dislocation loops following fission fragment irradiation might be the channelling of a large number of displaced atoms along the $\langle 10\bar{1}1\rangle$ directions, followed by their coagulation and interaction with the lattice. Similarly, channelling between the basal plane sandwiches might fulfill the same function, although with a much reduced efficiency. These are both mechanisms by which energy and momentum carried with point defects could be transmitted for long distances in the lattice, and could be a factor in explaining the observation that the visible damage in thick crystals of lead iodide is quite general, and is not localized to the initial path of the charged particle.

References

BRINKMAN, J. A. (1960), AEC Research and Development report NAA-SR-6642.

CHADDERTON, L. T. (1965), *Radiation Damage in Crystals*, Methuen (London).

CHADDERTON, L. T., and TORRENS, I. MCC. (1966), *Proc. Roy. Soc. A* **294**, 93.

GAY, W. L., and HARRISON, D. E., JR. (1964), *Phys. Rev.* **135**, A1870.

GIBSON, J. B., GOLAND, A. N., MILGRAM, M., and VINEYARD, G. H. (1960), *Phys. Rev.* **120**, 1229.

ROBINSON, M. T., and OEN, R. S. (1963), *Appl. Phys. Lett.* **2**, 30.

SILSBEE, R. H. (1957), *J. Appl. Phys.* **28**, 1246.

CHAPTER ELEVEN

Atomic Displacement in Alkali Halides

We now see how the technique of computer simulation may be used to give us a more detailed insight, at least qualitatively, into the phenomena which occur on an atomic scale when a crystal is disturbed by the passage of a fission fragment. The difficulties involved in drawing conclusions of a more quantitative nature from the results have been noted. Apart from the lack of knowledge of the interatomic potential in the dynamic energy range, the non-central forces associated with covalent crystal binding in the lead iodide lattice limited the accuracy of the computations in that compound. In order for computer simulations to be quantitatively accurate, we should require a precisely determined interatomic potential over a large range of interaction distances, the crystal atoms or ions must interact with purely central forces, the computational time-step must be very small, and either the boundary conditions should be exceptionally good or the computer must be larger and faster than any available at the present time. We can, however, draw reasonable and significant conclusions from simulations carried out under much less stringent conditions, if we choose a more simple lattice with a better known interatomic potential.

We propose in this chapter to describe the results of some computer simulations of ionic motion in three-dimensional potassium and sodium chloride crystallites (Torrens and Chadderton, 1967). Some of these results will further demonstrate very effectively the usefulness of the computer simulation technique in predicting effects which, by means of the necessity to take into account many-body collision processes, would have been difficult to predict by analytical means. The dynamics of ionic motion in face-centred cubic ionic crystals such as the alkali halides is of particular interest since many theories of colour centre defect formation involve the motion and displacement of lattice ions following irradiation, and since the lattice is pseudo-simple-cubic, so that new directional effects should be in evidence. The binding, moreover, is ionic and the interatomic potential contains a long-range electrostatic contribution in addition to the short-range closed shell repulsion. The discrepancy between ionic sizes and masses in an alkali halide should also affect the energy loss from a primary event, and have a marked influence on focusing phenomena.

Computational Procedure

The structures which will concern us here are those of potassium chloride, in which the alkali and halide ions are similar in size, and sodium chloride for which there is an approximately two-to-one ratio in ionic size. Both structures are strictly face-centred cubic but may be 'constructed' by arranging alternately the anions and cations on the lattice points of a simple cubic lattice. Potassium chloride has a lattice parameter of 3·14 Å and the sodium salt has fundamental cell dimensions of 2·81 Å. Some principal crystallographic directions in a potassium chloride lattice are shown in Figure 77.

A face-centred cubic section of an infinite crystal was simulated in the machine and one ion was given a selected kinetic energy, as though struck by an incoming fragment. The computer then employed iteration procedures similar to those described in the previous chapter to find the positions and

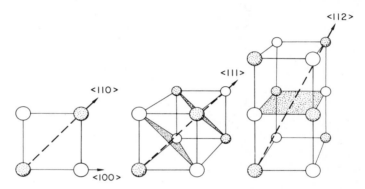

Figure 77. Principal crystallographic directions in potassium chloride. Shaded areas show 'ionic lenses' through which focusons in ⟨111⟩ and ⟨112⟩ directions must pass.

velocities of the ions of the crystallite at the end of consecutive equal intervals of time.

For convenience in calculation and in comparative analysis of the results, fundamental units of similar significance to those described in Chapter 10 were selected. These were slightly different for potassium and sodium chlorides, and are shown in Table VIII.

The unit of length was the equilibrium lattice spacing of the compound under consideration and the unit of energy was 1 eV in both cases. The unit of time was such that an alkali ion with energy of 1000 eV had unit velocity. Units of mass and velocity followed from these definitions, and the mass unit was such that the alkali ion in both cases had a mass of 2000 in computational units.

The crystallite of the simulation contained 1000 ions in a 10 × 10 × 10 cubic array with {100} faces. The input routine allowed any number of ions to be set moving initially with stipulated energies, and ions could be placed in non-lattice positions. Since only elastic energy losses were considered, and no electronic interactions were allowed, the initial energy in all cases did not exceed 1000 eV. In this region, electronic energy losses were assumed to be small enough to be neglected.

The Interionic Potentials

For an ionic crystal, it is a basic assumption that the solid may be considered as a system of positive and negative ions, the force between two such ions depending only on their distance apart, and having no directional variation. This interaction has two components: the electrostatic force which is repulsive or attractive depending on whether the interacting ions are of the same or opposite sign of charge, and the repulsive force due to the overlap of the closed electron shells of neighbouring ions.

TABLE VIII

Computational units for potassium and sodium chlorides

Unit	KCl	NaCl
Length	3.14×10^{-8} cm	2.81×10^{-8} cm
Time	4.47×10^{-15} sec	3.07×10^{-15} sec
Mass	3.25×10^{-26} g	1.91×10^{-26} g
Energy	For both 1.602×10^{-12} erg = 1 eV	
Velocity	7.025×10^{6} cm/sec	9.16×10^{6} cm/sec

For a two-body interaction, two main forms of repulsive potential have been used (Pauling, 1939; Born and Mayer, 1933; Verwey, 1946; Fumi and Tosi, 1964):

$$V(r)_{repulsive} = \lambda/r^n \qquad (11.1)$$

$$V(r)_{repulsive} = A \exp(-Br) \qquad (11.2)$$

The inverse nth power potential is empirical, while the Born-Mayer exponential form is based on a calculated statistical electron density distribution in the ion. The former increases rapidly to infinity with decreasing values of r, whereas the Born-Mayer exponential approaches a finite limit as r approaches zero (Figure 78). In the region of r in which these potentials are more accurate, near the equilibrium position, they are both sufficiently small in comparison to the electrostatic term for their difference to be insignificant. The inverse nth power potential is, in our view, the more accurate in the region of interaction which is of most interest in dynamic events (up to a few hundred electron volts). Events run with both potentials in this energy regime produce results which we feel substantiate this conclusion. There are two generally accepted methods of assigning a value to the index n. Some investigators have used the potential which has become known as the Born-Mayer-Verwey potential, following a treatment by Verwey (1946) of the ionic shell repulsion in the molecule. In this, a value of 12 is derived for the index n. This, however, is somewhat too hard for dynamic motion of the ions, where the interionic distance is often considerably less than the equilibrium distance in the crystal. In addition, it emerges from a consideration of the repulsive interaction in a single molecule, rather than in the crystal. A more reasonable potential for our purpose is the Pauling form (1939), where the value of n depends on the closed shell ion size, being 9 for K^+–K^+, K^+–Cl^-, and Cl^-–Cl^- interactions, 8 for Na^+–Cl^-, and 7 for Na^+–Na^+. Thus, the potential is slightly softer for the smaller ions. The value of λ in each case was found from a consideration of the equilibrium ionic configuration in the crystal. However, in the case of NaCl, the same constant $\lambda_{Na^+–Cl^-}$ (being the one found from equilibrium data) should not be used for Na^+–Na^+ and Cl^-–Cl^- interactions, since the repulsive force between these two pairs of ions

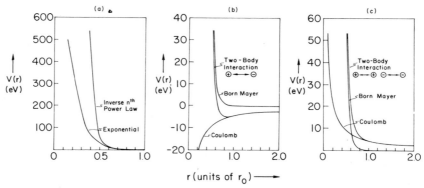

Figure 78. Interionic potentials in potassium chloride. (a) Comparison of Born-Mayer exponential and inverse nth power form of the repulsive potential; (b) and (c), computational potentials in KCl, with their repulsive and Coulomb components.

at the lattice separation is considerably different from that between sodium and chlorine ions. Accordingly, the potential constants for the identical ion interactions were changed so that the repulsive energy at the equilibrium lattice separation was different from that of sodium and chlorine ions by the eighth power of the ratio of the respective sums of the ionic radii:

$$\lambda_{Na^+ –Na^+} = \lambda_{Na^+–Cl^-} \left(\frac{2r_{Na^+}}{r_{Na^+} + r_{Cl^-}} \right)^8 \tag{11.3}$$

$$\lambda_{Cl^- –Cl^-} = \lambda_{Na^+–Cl^-} \left(\frac{2r_{Cl^-}}{r_{Na^+} + r_{Cl^-}} \right)^8 \tag{11.4}$$

The electrostatic form of the potential was of the form:

$$V(r)_{electrostatic} = e_{eff}^2 / (\epsilon r) \tag{11.5}$$

where e_{eff}^2/e^2 was 0·5 for KCl and 0·64 for NaCl, e being the electronic charge. The ϵ represented a dielectric reduction which was 1 for interaction with nearest neighbours, and 2 for next nearest neighbours and greater (Szigeti, 1949). These reductions to the basic Coulomb potential made some allowance for polarization of the ions, and for electronic screening.

The final potentials used in the computation were thus:

KCl
$$V(r)_{K^+-K^+,\ K^+-Cl^-,\ Cl^--Cl^-} = \frac{0·135}{r^9} \pm \frac{4·585}{\epsilon r} \qquad (11.6)$$

NaCl
$$V(r)_{Na^+-Na^+} = \frac{7·11 \times 10^{-3}}{r^7} + \frac{5·123}{\epsilon r} \qquad (11.7)$$

$$V(r)_{Na^+-Cl^-} = \frac{0·163}{r^8} - \frac{5·123}{\epsilon r} \qquad (11.8)$$

$$V(r)_{Cl^--Cl^-} = \frac{1·23}{r^9} + \frac{5·123}{\epsilon r} \qquad (11.9)$$

where r is in computational units (Figure 78). Rigorously, the values of λ for the different ionic interactions in KCl are slightly different, but since the ionic radii of potassium and chlorine ions are very similar, the approximation is considered reasonable. The electrostatic term of the potential is considerably larger than the ion shell repulsion at distances in the region of and greater than the equilibrium separation of the ions in the crystal, and is thus more influential for long-range interactions.

It is important at this point to stress again that in utilizing the computer simulation technique we are not seeking accurate quantitative results as regards the motions of individual ions, nor yet concerning the energy spread from the origin of a primary event. This is a pseudo-experimental technique. We hope that, by obtaining a qualitative or semi-quantitative picture of what happens in a crystal when one ion receives a certain energy, we may be in a better position to understand those phenomena which are related to ionic motions in crystals such as the alkali halides, than we should be by either analytical calculation or simple empirical considerations alone. We are also of the opinion that the potentials which have been employed are reasonable; first of all since they are based on theoretical knowledge of binding in ionic crystals; and secondly since the effects of altering them within reasonable limits, and of trying different accepted analytical forms, are such as to alter the detailed ionic trajectories only slightly without having more than a trivial effect on the general progress of a damage event.

Another observation which should be clearly stated here is that, in a perfect crystal, the effect of completely cutting off the potential between two ions at a separation of $3r_0$ (r_0 = equilibrium (100) lattice spacing) is negligible for events in the dynamic energy range. Again, detailed numerical results are altered, but the general conclusions for a particular event remain substantially the same.

Results

Focusing in the {100} plane

Interpretation of computer simulation results is fraught, as we have discovered, with many pitfalls for the unwary, and proper care must be exerted in the extrapolation of results from the model to the solid. We shall see later, however, that the $\langle 110 \rangle$ direction plays a dominant part in many low energy displacement and sub-displacement events, and for this reason we present in this section some discussion of events initiated entirely in a {100} plane. It is clear, of course, that an ion moving in such a plane may well be in a state of unstable equilibrium. Only a small component of motion out of the plane – easily supplied by the vibrating lattice – may be necessary to send the ion into the third dimension and to absorb the excitation more rapidly. Yet the behaviour of ions neighbouring the knock-on, the efficiency of focusing down the like-ionic $\langle 110 \rangle$ directions, and the observation that an allowed relaxation of neighbouring {100} planes does not greatly alter the course of a typical event, prompt us to present the results in the way we have chosen.

Ionic Lens Anomalous Effects

The results of some events run in potassium chloride are shown in Figures 79 to 82. In the event depicted in Figure 79 a potassium ion was given an energy of 80 eV and was projected in a direction making an angle of $\tan^{-1}(3/2)$ with a $\langle 100 \rangle$ direction (that is at about $11 \cdot 3°$ to a $\langle 110 \rangle$ line). This resulted in a strongly assisted focusing of energy down the $\langle 110 \rangle$ line, in which each succeeding ion of the sequence was deflected by a single 'ionic lens' consisting of two chlorine ions, so that it subsequently made an almost head-on collision with the next ion in the line. Approximately 55 % of the initial energy is carried out of the array in this way, in a time of 10^{-13} sec, before many of the ions in the vicinity of the relevant $\langle 110 \rangle$ line have begun to respond to the disturbance. The continuation of the trajectories during the relaxation period revealed some characteristics of the replacement process and of the effect of the ionic lens. Thus, while the initial ion, following its collision with the next one in the $\langle 110 \rangle$ line, succeeded in crossing back through the potential barrier and eventually in returning to its original position in the lattice, the subsequent ions in the focusing line were forced to replace sequentially. The 'simple lens' analogy will not predict this behaviour. The ion lens must be permitted to move and vary in strength under the influence of the passing ion, becoming astigmatic with a time-dependent focal length. Thus, the initially knocked-on ion, when it was reflected back from its collision with the next ion in the $\langle 110 \rangle$ line, found that the assisting lens, after a temporary disturbance, had reformed in a position where it aided a return to the original lattice site. For succeeding ions in the focusing line, the reconstituted potential barriers were in a position to force the ions (which had small velocities after their principal head-on collisions) into replacement sites. It is suggested, therefore, that the reason for the anomalous behaviour of the knock-on is the fact that it received its energy

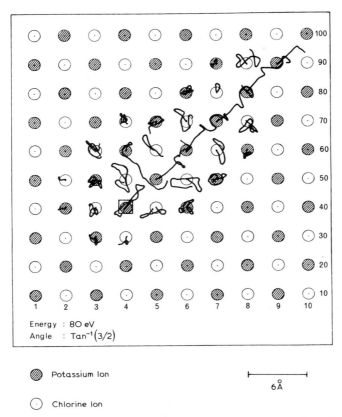

Figure 79. Progress of a displacement event in which a potassium ion (boxed) receives an energy of 80 eV and moves off at about 11·3° to [100]. Focusing occurs strongly in the [110] direction, removing over half of the initial energy from the array. A sequence of replacements occurs along the line, and a possible final configuration is an interstitial somewhere outside the array, and a vacancy in the immediate vicinity of the primary event.

instantaneously, on its lattice site, while subsequent ions were *gradually* accelerated from rest during the collision time. The resulting difference in the disturbance of the ions of the lens was sufficient to cause permanent displacement in the secondary cases, and the return of the primary to its lattice site. This type of damage trajectory pattern has been observed in a large number of events with varying initial energy and angle of projection of the primary ion.

Another manifestation of an anomalous lens effect may be seen in the probable final configuration of this event. Computer time limitations rendered it impracticable to continue the run until the ions settled down completely, but it is possible to be reasonably certain of the eventual defect pattern. It seems that a vacancy will be left, not in the ⟨110⟩ line of the focusing sequence, but in a neighbouring ⟨110⟩ potassium line, the ion diffusing from its site through a temporarily absent barrier into the line of the focusing sequence. This result and the anomalous failure to replace of the primary knock-on are symptomatic of the difference between this type of computation

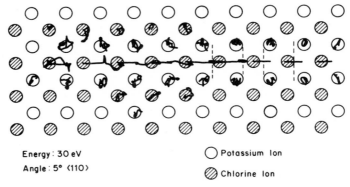

Energy: 30 eV
Angle: 5° ⟨110⟩

◯ Potassium Ion
◍ Chlorine Ion

Figure 80. Degradation of a dynamic into a static crowdion. The potassium ion receives an energy of 30 eV at 5° to [110], resulting in the production of a symmetrical crowdion interstitial on the [110] line six lattice spacings away. A perfect focuson travels on out of the set.

and an analytical calculation of a focusing sequence, which would find difficulty in predicting the influence of the ionic lenses during the relaxation period.

The ⟨110⟩ dynamic crowdion of this event will in all probability be terminated by an interstitial crowdion configuration after a number of further focusing collisions. This type of crowdion configuration has been shown to be stable in a separate run (Figure 80). For this event a potassium ion in a large array was set moving with an energy of 30 eV at 5° to the ⟨110⟩ line. The final configuration consisted of a vacancy in the place of the second ion, with a stable ⟨110⟩ crowdion some six ⟨110⟩ lattice distances from the original knock-on. The energies were not sufficient for the ion from the neighbouring potassium line to find its way through the barrier into the focusing line. During the propagation of the dynamic crowdion along this line, the ions in the immediately neighbouring ⟨110⟩ chlorine rows were observed to relax in such a way that a wave of charge polarization followed the propagation of the sequence through the lattice. This effect was particularly noticeable during low energy events. The configuration of the static crowdion after the moving ions had relaxed into their stable positions was symmetrical with respect to the lattice site, and the neighbours in the ⟨110⟩ line came to equilibrium in positions displaced outward along ⟨110⟩ from the original lattice sites.

An event identical to that of Figure 79, with the exception that the initial potassium ion energy has been raised to 130 eV, is depicted in Figure 81. In this case the energy was sufficiently high for the first ion to replace, since it still retained forward momentum after its initial collision with the second ion in the ⟨110⟩ line. Also of note in this event are the secondary ⟨110⟩ chlorine ion focusing chains at 90° to the main sequence. These were not sufficiently

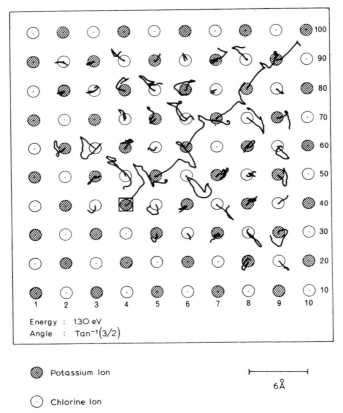

Energy : 130 eV
Angle : Tan⁻¹(3/2)

⊘ Potassium Ion

⊙ Chlorine Ion

6 Å

Figure 81. An event similar to that of Figure 79, with the initial energy raised to 130 eV. Now the first ion also is displaced, and we note that secondary [110] focusing is commencing at 90° to the main sequence.

energetic for replacements to occur but were nevertheless influential in efficiently removing energy to considerable distances from the principal focusing line.

The course of a further event involving an 80 eV potassium ion, this time travelling initially at 10° to ⟨100⟩, is portrayed in Figure 82. The energy became focused along ⟨100⟩, but through lack of any assistance from adjacent lines the efficiency was considerably less, and only 29% of the primary kinetic energy passed out of the array as a focuson, in a time of 7×10^{-14} sec. The region of disturbance in the lattice extended further from the $\langle 100 \rangle$ line in question, covering about nine lattice spacings, compared to approximately four in the event of Figure 79. Replacement of unlike ions in $\langle 100 \rangle$ focusing was not expected and did not occur at energies of this order, owing to the absence of an assisting potential barrier.

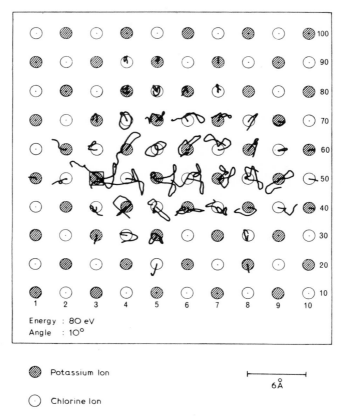

Figure 82. The initial direction of motion of the struck potassium ion is in this case at $10°$ to [100], illustrating the relatively inefficient simple focusing process. No [100] replacement occurs, being energetically unfavourable in lines of alternately oppositely charged ions.

The dynamic motion of ions in the *sodium chloride* lattice was influenced by the discrepancy in ion size of the sodium and chlorine ions. The focusing efficiency was consequently less, and, for events with identical initial conditions to that of Figure 79 in KCl, the energy contained in the $\langle 110 \rangle$ focusing line after four lattice spacings was, for both sodium and chlorine chains, less than 50% of the initial energy. The replacement characteristics were also somewhat different to those for KCl. Figure 83 shows the $\langle 110 \rangle$ focusing in events where sodium and chlorine ions were projected, again at 11·3° to ⟨110⟩, with 80 eV of kinetic energy. We see that, in the event where the chlorine ion was initially projected, the replacement pattern is similar to that for KCl, although in this case the first ion, as well as subsequent ions of the ⟨110⟩ line, took part in the replacement sequence. Sodium ion focusing appeared different in that the effect of collisions with the large chlorine ion lenses was sufficiently severe to cause rapid defocusing in the original ⟨110⟩ line, although subsidiary focusons

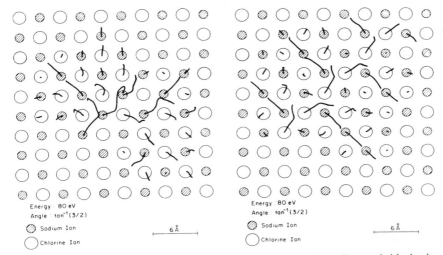

Energy 80 eV
Angle tan⁻¹(3/2)

⊘ Sodium Ion   ⊢——6Å——⊣
◯ Chlorine Ion

Energy 80 eV
Angle tan⁻¹(3/2)

⊘ Sodium Ion   ⊢——6Å——⊣
◯ Chlorine Ion

Figure 83. Events with the initial conditions of that of Figure 79 for sodium and chlorine ions in sodium chloride. Focusing is less efficient because of the size discrepancy of the ions of this lattice. Subsidiary [110] focusons are generated in the chlorine ion event, and an interesting effect in the sodium ion focuson is its removal through interaction with a chlorine ion lens to the neighbouring parallel sodium ion [110] line.

developed in other ⟨110⟩ sodium lines. When the ionic motion was almost directly along ⟨110⟩, subsequent energy losses of the focuson to the lenses were low owing to the small size of the sodium ions. In the chlorine ion focusing sequence, each ion was only slightly deflected inwards by the sodium ion lens, with the result that the initial direction of motion of successive ions approached the ⟨110⟩ direction only very slowly.

Focuson Propagation and Energy Loss

The consequences of the differences in trajectory of identical events run in NaCl with sodium and chlorine ion focusing and in KCl with potassium ion focusing may be illustrated by reference to plots of focuson energy as a function of time for the three runs. Thus, in Figure 84 the maxima of the kinetic energy curves for successive ions of the focusing line are considered to be representative of the transmitted focuson energy. We see that the sodium ion focuson initially loses energy quite rapidly, and after two collisions (when the ions of the line are travelling almost directly along $\langle 110 \rangle$) its rate of energy loss decreases to less than 2 eV per collision. On this basis the disturbance might be expected to propagate some fifteen or twenty $\langle 110 \rangle$ lattice distances. The chlorine ion focuson loses energy at a more constant rate (some 5 to 8 eV per collision), although the rate decreases slightly as the energy drops. Its range is therefore between five and ten $\langle 110 \rangle$ units. The picture for potassium ion focusing in KCl is very similar.

The reason for the differences in focuson propagation in the three cases is basically another effect of the lenses through which the focused ions have to pass. If we compare the shapes of the uppermost parts of the energy curves for

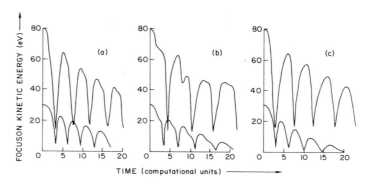

Figure 84. Kinetic energy curves for ions of 80 and 30 eV, $\langle 110 \rangle$ focusons: (a) K$^+$ in KCl; (b) Na$^+$ in NaCl; and (c) Cl$^-$ in NaCl. An initial sharp drop in focuson energy occurs over the first one or two collisions, followed by a more constant rate of energy loss as the ionic trajectories approach the axis of focusing.

sodium with those for chlorine in NaCl and potassium in KCl, it becomes clear that the chlorine ions experience very little detectable influence of the lenses, whereas the others, in differing degrees, are sensitive to the barriers. This result may be confirmed by reference to equivalent ion energy curves for motion directly along $\langle 110 \rangle$ in the three cases. In Figure 84, when the initial ion energy is 30 eV (reduced to increase the sensitivity to the lens action), for potassium in KCl and chlorine in NaCl no kink in the energy peak is detectable, the collision with the next ion occurring before any acceleration after passing through the lens could be experienced.

The general pattern for focuson propagation in the $\langle 110 \rangle$ direction is therefore an initial drop in energy over the first one or two collisions, being the more pronounced the further off-axis the initial direction of motion, followed by an almost constant energy loss as the focuson is attenuated, though the rate of energy loss does in fact decrease slightly as the energy drops, even where the initial projection is directly along $\langle 110 \rangle$ (Figure 84).

It should be stressed that in all of these plots the ions have not come permanently to rest when their energy curves are discontinued. Diffusive relaxation motion has been neglected since it has no major effect on the propagation of the focuson.

Energy Correlation Diagrams for Alkali Halides

The 80 eV correlation diagram for a struck potassium ion in KCl is illustrated in Figure 85, and the relative importance of certain crystallographic directions in the focusing of energy away from a primary event becomes immediately evident. In particular, the ⟨110⟩ assisted focusing is extremely predominant

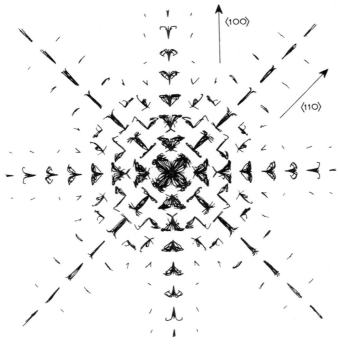

Figure 85. 80 eV energy-correlation diagram for a struck potassium ion in a {100} plane of potassium chloride. The diagram is built up from the superimposed trajectories of events in which the initial direction of motion is varied through 360°. The predominance of ⟨110⟩ focusing and the effect of the ionic lenses is clearly illustrated.

and much more intense than the ⟨100⟩ focusing. The main ⟨110⟩ lines carry some energy from almost all of the events, and a large fraction of the energy from some initial projections close to the ⟨110⟩ directions, while in the ⟨100⟩ case only events where the ion is set moving at under 20° to the line focus energy over more than about three lattice spacings. Subsidiary ⟨110⟩ focusing in adjacent lines is beginning to develop, and, if the events were permitted to continue into the relaxation period, this would undoubtedly become more marked.

Some other facts come to light through this type of orbit diagram:

(1) The size of the affected region is governed by the energy of the knock-on, and the disturbance is generally concentrated along or near to the principal focusing axes.

(2) The disturbance of the $\langle 110 \rangle$ lenses by the traversal of the focuson is clearly illustrated.

(3) The fact that the lattice is tightly bound is made evident by the absence of trajectories of neighbouring ions which touch or cross. If the events which make up this picture were run into the relaxation period, some trajectories would cross during $\langle 110 \rangle$ replacement and diffusive motion of the ions. This

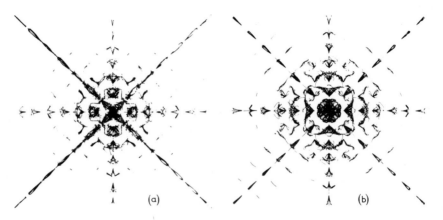

Figure 86. Energy-correlation diagrams for 80 eV primary knock-ons in sodium chloride: (a) sodium ion, (b) chlorine ion. Much more overlapping of trajectories occurs in the sodium ion case, and a greater disturbance of the lenses characterizes the chlorine ion diagram.

should be compared with the many crossing trajectories of the 80 eV lead iodide correlation diagram (Chapter 10, Figure 74).

(4) The symmetry of the spread of the damage region is very pronounced.

Similar correlation diagrams for sodium and chlorine ion primaries in NaCl are shown in Figure 86, though the computation time for these was larger than in Figure 85. The effect of the different ion size is immediately obvious. Much more overlapping of trajectories occurs in the sodium case, the centre of the pattern having only small 'forbidden areas' free of trajectories. Also noticeable is the greater disturbance of the lenses for the chlorine primary, and the higher degree of focusing in the sodium $\langle 110 \rangle$ sequences.

Focusing in the $\langle 111 \rangle$ Direction

The $\langle 111 \rangle$ direction in a cubic crystal is interesting from the point of view of focusing, since between each collision there are two triangular lenses of ions to be traversed (Figure 77). In an alkali halide, moreover, the direction is particularly interesting since the ions in the focusing line are alternately alkali and halide ions, and the two assisting lenses each consist of three ions of identical type. This affords an opportunity to study the transmission of energy through focusing in a doubly assisted chain.

Investigation of ⟨111⟩ focusing was carried out for different initial energies and directions of motion as in the {100} plane. As might be expected from the lens geometry, assisted focusing is very pronounced and should lead to a higher focusing efficiency in this direction. However, the advantages in low energy events were counteracted by energy losses to the lenses. We remember that there are two of these in each focusing unit, and we note that in each lens there are three ions, all at only slightly greater distances from the axis than the two

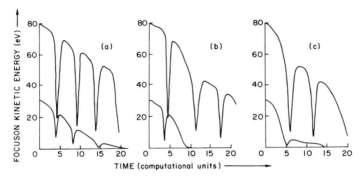

Figure 87. Kinetic energy curves for the ions in 30 eV and 80 eV axial ⟨111⟩ focusons: (a) K⁺ in KCl, (b) Na⁺ in NaCl, (c) Cl⁻ in NaCl. Much more energy is lost to the lens ions than in the ⟨110⟩ case, and in NaCl the ion size difference along the ⟨111⟩ line further reduces the focusing efficiency.

most important ions of the ⟨110⟩ lens. Thus, although ions moving initially off-axis were quickly reflected, the energy losses of even axially moving focusons to the lenses were considerable.

To investigate these effects axial 30 eV and 80 eV focusons were initiated and the ion energy curves were plotted (Figure 87) for comparison with ⟨110⟩ focusing (Figure 84). In all cases (K⁺ in KCl, Na⁺ and Cl⁻ in NaCl) the range was much less than that for a 30 eV focuson in ⟨110⟩ {100}. The curves reveal very closely the action of the lenses, the critical factor being the amount of energy lost by one ion of the sequence before the next ion begins to move (i.e. before the commencement of the next focusing collision). Potassium and chlorine ions initiating the sequence in KCl were almost identical and so only one is shown here. In NaCl the chlorine ion, being the largest, lost a great deal of energy to the lenses (one of which is itself composed of chlorine ions). So when a chlorine ion of 30 eV started the sequence, the next ion in the ⟨111⟩ line, a sodium, received only about 5 eV maximum kinetic energy and the disturbance failed to penetrate the next chlorine lens. If, on the other hand, a sodium ion initiated the focuson, its small size enabled it to penetrate both lenses with comparatively little energy loss, and sufficiently to pass more than 20 eV of its energy to the chlorine ion. The latter, however, then experienced the same energy transfer difficulty as before and the focuson died there. A further significant point was the observation that the energy lost to the two lenses was generally funnelled off in both structures down ⟨110⟩ directions as long-range focusons, thus confirming the vital importance of ⟨110⟩ focusing in these lattices.

Events run at 80 eV showed the same general pattern, the ion size discrepancy

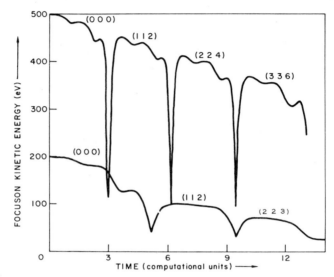

Figure 88. Kinetic energy for 'impulse' focusing in the [112] direction, at 200 and 500 eV initial energy. The former survives only one focusing collision, degenerating into a [111] focuson, but as the energy is increased the trajectory is only slightly modified by collisions with the asymmetric lens, and the focuson survives several collisions. The effect of collisions with the lens ions is demonstrated by the shape of the energy maxima of the separate ions. Numbers in brackets refer the position co-ordinates of ions whose energy is plotted to the initially struck ion as origin.

in the ⟨111⟩ line of NaCl, and in the assisting lenses, causing much more rapid attentuation of the focuson than in KCl. Further variation of the focuson energy, up to 500 eV in KCl, confirmed that ⟨111⟩ is indeed a good focusing direction in this compound, the energy loss per focusing collision distance being approximately 10 eV, for ions travelling directly along the axis. In the case of non-axial focuson propagation the energy loss was greater, but the symmetry and geometry of the assisting ionic lenses was such that the solid angle for projection of an ion for focusing of energy to occur was quite large.

Higher-Order Focusing – The $\langle 112 \rangle$ Direction

The $\langle 112 \rangle$ direction and other higher index directions in the crystal are some-what different from those which we have so far discussed. The assisting poten-tial barrier is in the $\langle 112 \rangle$ case an asymmetric lens consisting of four ions of a {100} plane, and the distance between focusing collisions is some 2·45 lattice spacings. Consequently, we expect much more critical conditions for the initial energy and direction of motion to prevail if focusing is to occur.

This is indeed the case for normal focusing energies. However, events run in this direction have demonstrated a new concept of assisted focusing in which the phenomenon is rendered possible by an *impulse* type of situation similar to that which is assumed in analytical momentum approximation calculations. For initial energies up to over 100 eV, no focusing occurs, and the trajectory of the first ion travelling in the $\langle 112 \rangle$ direction is so modified by collision with its neighbours that it collides almost head-on with a $\langle 111 \rangle$ neighbour. For a 200 eV ion, less deviation from the initial trajectory results from the collision with the neighbours, and less energy is given to the $\langle 111 \rangle$ ion. The initial ion then proceeds to collide with its neighbour in the $\langle 112 \rangle$ direction, which receives 100 eV. In Figure 88 we have plotted the kinetic energy of the relevant ions against time, thereby illustrating the sequence of collisions and their effect on the focuson.

If we now increase the initial energy to 500 eV, we see very clearly the impulse effect as a greatly enhanced $\langle 112 \rangle$ focusing (Figure 88). The trajectory of the 500 eV ion is scarcely affected by collision with its $\langle 001 \rangle$ and $\langle 111 \rangle$ neighbours, and it goes on to give up almost all of its energy to the next ion of the $\langle 112 \rangle$ line. The focuson energy loss in this case is less than 50 eV per focusing collision. The shapes of the kinetic energy peaks for each focusing ion show the relatively minor effect of the two collisions with lens ions which occur between $\langle 112 \rangle$ collisions. Furthermore, at higher initial energies we should expect even less energy loss per collision. However, the asymmetry of the lens has one major disadvantage even at these energies, as may be demonstrated by considering an initial direction of motion which is slightly off the $\langle 112 \rangle$ axis, for example by an angle of about 7°. In this case the focuson is rapidly attenuated.

Thus, events run in the $\langle 112 \rangle$ direction reveal a *lower* energy limit for focus-ing, dependent not, as in the $\langle 111 \rangle$ focusing, on the ability of the energetic ion to penetrate the assisting lens, but on the ability of the ions of an asymmetric lens to modify the trajectory of the traversing ion, thereby terminating the focuson. It is perhaps more correct to call this particular phenomenon 'hindered' rather than assisted focusing, and it is the opinion of the authors that, in view of the rather critical conditions required for the direction of motion of the first ion, *impulse* focusing in this direction in KCl-type alkali halide lattices will necessarily be inefficient. The same argument of course applies to higher-order focusing.

Impulse focusing proceeds in a similar way in the lattice of the sodium salt for both sodium and chlorine ions, and once again there is an effective lower energy limit for the process below which focusing is effectively hindered. In the case of the sodium ion, however, there are other possibilities which arise out of the comparatively small size of this ion. As a typical example, let us consider the case in which a sodium ion is set moving in the $\langle 112 \rangle$ direction with an energy of 80 eV. As a result of a collision with a neighbouring chlorine ion the sodium knock-on becomes deflected in such a way that it moves into a field of low interaction potential between adjacent $\langle 100 \rangle$ rows of ions, becoming effectively axially channelled. Simultaneously a $\langle 111 \rangle$ chlorine focuson initiated by the first collision rapidly degenerates into a $\langle 110 \rangle$ focuson, and

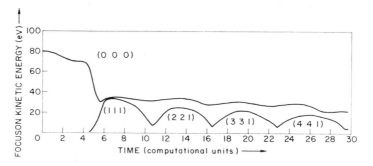

Figure 89. A sodium ion set moving with energy 80 eV in the [112] direction becomes channelled between [001] lines of ions as a result of its collision with the nearest neighbour in the [111] direction. The energy/time curve shows the resulting simultaneous propagation of the channelon and a [110] focuson. Numbers in brackets have similar significance to those of Figure 88.

both focuson and channelon persist until the termination of the computation of the event (Figure 89). This once again indicates the relative importance of the $\langle 110 \rangle$ axes as fundamental directions for energy dissipation by focusing mechanisms, and confirms that in diatomic solids in which one atomic species is considerably smaller than the other, there is a finite probability of the far transfer of both mass and energy from the primary event.

The Ionic Displacement Threshold

Many of the primary damaging events which were simulated culminated frequently in defect configurations which typically comprised several ionic replacements to substitutional sites and one permanent *displacement* to an interstitial site. It was logical, therefore, to ask how the displacement threshold E_d, the minimum knock-on energy required for the introduction of a permanent Frenkel pair, varied with the initial direction of projection. In particular, since there is a marked effect of assisting ionic lenses in $\langle 110 \rangle$ displacements events in a {100} plane, we were interested in determining what effect this might have on the angular variation of E_d.

In order to establish the angular dependence of the variations of the displacement threshold in the {100} plane potassium knock-ons in KCl moving with various energies at initial angles varying from 0 to 45° with $\langle 100 \rangle$ were investigated. In this manner the computational threshold for the displacement of an ion (not necessarily the primary) was determined to within a few electron volts, and was seen to vary from a minimum for primary motion along $\langle 110 \rangle$ to an indeterminate maximum in the $\langle 100 \rangle$ direction.† Similar computations

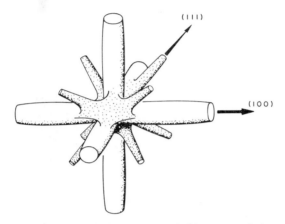

Figure 90. Schematic polar plot of displacement threshold energy variation with direction of motion of an ion in the potassium chloride lattice. Indeterminate maxima occur in the $\langle 100 \rangle$ and $\langle 111 \rangle$ directions, decreasing to 25 to 30 eV minima in $\langle 110 \rangle$ directions.

were performed for initial directions of motion not contained in the {100} plane, and a schematic polar plot for displacement energies in the KCl lattice is given in Figure 90. The displacement threshold for a chlorine knock-on had almost identical variations owing to the similarity in size of the two ions and owing to the high symmetry of the lattice.

It therefore appeared that the displacement threshold for potassium in the $\langle 110 \rangle$ direction is between 25 and 30 eV, and this conclusion was substantiated and the value roughly checked numerically by reducing the magnitude of the computational time-step and repeating the run. For energies greater than 30 eV the ions of the lenses were always sufficiently perturbed to move out and be reflected by *their* barrier lenses, whence they returned to prevent the ion of the primary sequence from rejoining its lattice site after making a focusing collision.

† No subsidiary maximum similar to that seen by Erginsoy, Vineyard, and Englert (1964) was observed in the $\langle 110 \rangle$ direction. If present it may be concealed by the inherent error in our energy estimates.

13

For lower energies the primary lens was only disturbed sufficiently to allow the diffusion of the ion of the sequence back to its original site, but not enough for the two lens ions to interact strongly with their own neighbours. At very low energies the initial ion was, of course, unable even to penetrate the primary barrier lens.

The threshold energy for displacement of a potassium ion directly along the $\langle 100 \rangle$ direction has not been accurately determined to lie within certain limits but is undoubtedly in excess of 150 eV, and is probably nearer to a value in the neighbourhood of 200 eV. Uncertainty in this value derives from the long computational time which proved necessary to follow a typical event, beginning with an energetic impulse along or near to $\langle 100 \rangle$, to conclusion. The displacement process, moreover, was essentially different from that which normally took place in or near to $\langle 110 \rangle$, and was not accompanied by a string of sequential replacements. Indeed the computations indicated the probable permanent displacement of either the primary knock-on itself or its immediate neighbour, and a large consumption of energy in the generation of $\langle 110 \rangle$ focusons in the neighbouring lattice. Displacement along $\langle 111 \rangle$ displayed a similar pattern of behaviour with an indeterminate maximum threshold for ejection directly along the axis. In this case, however, the threshold falls away much more quickly with departure from axial projection – as the schematic representation of the three-dimensional displacement surface (Figure 90) indicates. The basic pattern of the energetic events indicated in all cases the critical rôle of the barrier ionic lenses in determining whether or not permanent displacement occurs. The kinetics of the lens ions in relation to those of the focusing ion, and what happens *after* the focusing collision has occurred are the influential factors once the energy for initial penetration of the barrier has been reached. The time variation of the time-dependent potential barrier, moreover, is extremely complicated, depending as it does on many-body effects, and would in all cases be impossible to predict or describe in analytical terms.

The displacement threshold for sodium in NaCl differed appreciably from that for chlorine. For replacement of the smaller sodium ions the assisting chlorine barriers were not greatly disturbed, and were therefore much more effective in preventing the return of the sodium ion to its lattice site. In addition sodium focusing collisions occurred much closer to the geometrical centres of the lattice sites on account of the smaller ionic size, thereby enhancing replacement and displacement. The numerical value for the displacement threshold along the all-important $\langle 110 \rangle$ direction, for example, was as small as 20 to 25 eV for sodium ions, and as large as 70 to 90 eV for chlorine.

The Effect of Lattice Vibrations

In the work described so far the simulated section of crystal lattice has in all instances been assumed to be at a temperature of $0°K$, and zero-point motion of the ions has been neglected. This of course is an idealized situation, and in

practice the lattice is vibrating in a system of quantized modes, whose characteristics depend on lattice dimensions and on temperature. The simulation of
realistic lattice vibrations in a calculation of the present nature would be
extremely difficult and time-consuming, but a qualitative approximation to the
physical picture may be undertaken.

The form of the approximation we have employed is similar to that first
used by Robinson and Oen (1963), and assumes that the time of collision of a

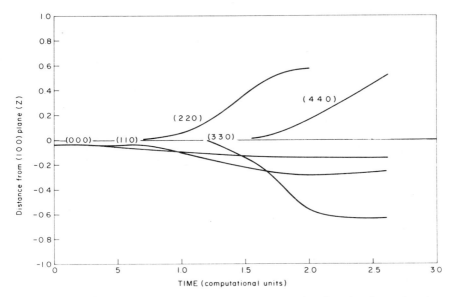

Figure 91. Plot of distance moved perpendicular to the (100) plane in units of r_0 against time
for ions of the [110] focusing line in an event with the identical initial conditions to that of
Figure 79, but including the finite-temperature approximation. A decrease in focused energy
and in range of the focuson is a consequence.

pair of ions is short in comparison to the vibrational period of a lattice ion.
The moving ion approaching a lattice ion assumes that the latter is not, as in
the zero-energy case, sitting on the lattice site, but at some position in the
neighbourhood. The choice of the position is made randomly by the computer
from a triangular neo-Gaussian probability curve, peaked at the lattice site,
and whose base size (the vibrational amplitude) and peak height are functions
of temperature. Thus the moving ion first 'sees' the stationary ion at a particular
stage of its vibration chosen at random. While this is reasonable as a first
approximation, it neglects the collective oscillations of the lattice ions, and
we do not therefore have a true physical picture. The approximation nevertheless indicates some of the consequences of the ions not being precisely placed
on their lattice sites.

One most obvious result which one might expect would be a decrease in the efficiency of focusing, owing to slight asymmetry of the lenses and the fact that it is impossible for a focuson to be carried directly along $\langle 110 \rangle$ or $\langle 111 \rangle$. Secondly, ions moving in a particular crystallographic plane will tend to be knocked out of the plane by collision with stationary ions. It was therefore decided to investigate the extent to which the pattern of damage events studied in the zero-vibration case would be affected. The event in which a potassium ion of KCl was given an energy of 80 eV at an angle of $\tan^{-1}(3/2)$ to $\langle 100 \rangle$, for example, was re-run with a thermal vibrational amplitude of $0 \cdot 1 r_0$, and the trajectories were plotted in the usual way. The difference between the zero and finite temperature cases was not very marked, with the exception that in the latter event the first ion in the sequence seemed to be remaining in the replacement position. The curves shown in Figure 91 are those of distance moved *out of the plane* (i.e. perpendicular to the {100} plane of projection) against time, for each ion of the $\langle 110 \rangle$ focusing sequence. Ions are at first ejected out of the plane and later reflected from the neighbouring planes. This naturally results in a decrease in the amount of energy focused down the $\langle 110 \rangle$ line and consequently in the range of a focuson. For example, in the 80 eV event in question, the focuson energy after five lattice spacings was $14 \cdot 5$ eV in the finite temperature case, as compared with an energy of 42 eV in the zero temperature event.

Discussion

One general result of these events in ionic crystals has been the relatively minor influence of the Coulomb contribution to the interionic potential. The use of different Coulomb terms has little effect on the progress of a damage event, although detailed ionic trajectories are slightly altered. The soft electrostatic potential is overshadowed, during the dynamic motions of the early part of an event, by the hard ion shell repulsion of the more energetic collisions. Only during the final relaxation of the lattice in regaining equilibrium does the Coulomb part of the interaction become the more influential. This of course means that the repulsive potential will play the major role in determining efficiency of focusing of energy and displacement thresholds. A related consequence of this is that the effect of cutting off the interionic potential at distances greater than $3r_0$ is negligible – a fact which is extremely useful from the point of view of computer time limitations.

Undoubtedly the most significant result of the computations has been the influence of the neighbouring ionic lenses during focusing. Analytical calculations on assisted focusing have permitted the ions of the lenses to relax with the passage of the focused ion, but have not allowed for subsequent motion of these ions or for their action on the focusing line during the relaxation period. Consequently, they cannot predict realistically the final configuration of the disturbed lattice, with particular emphasis on permanent displacement of the ions. We have noticed that, in the critical energy region for ionic displacement, below about 50 eV, the path followed by an ion moving in the $\langle 110 \rangle$ direction, for example, and where it is finally located are almost wholly dependent on its interaction with the relevant ionic lens following its focusing collision. Also, the energy limits between which focusing occurs in higher-order focusing are largely a consequence of the assisting lens geometry. To take this into account analytically would necessitate an immensely involved calculation, so that computer simulation offers an attractive approach to a more realistic model of lattice relaxation and dynamic motion.

The $\langle 110 \rangle$ replacement sequence in ionic crystals has attracted the attention of a number of workers in the field of colour centre formation, since it represents a feasible method for the removal of an ion from the immediate neighbourhood of a vacancy following irradiation, so that recombination cannot subsequently occur. It is certainly possible for a negative ion to be given several electron volts of energy, either directly or through some ionization or recombination process, by incident radiation.

One mechanism for F-centre formation by interstitial-vacancy separation which has been widely discussed does not directly involve $\langle 110 \rangle$ replacement. This process, suggested by Varley (1954), predicts the ejection of a multiple ionized (and therefore positively charged) halogen ion owing to its interaction with the six surrounding alkali ions. Controversy following this suggestion centred around the lifetime of the doubly ionized particle, since once the electron-hole recombination occurs the strong repulsion ceases, and the validity of the mechanism has been questioned, mainly from this standpoint of lifetime (Howard and Smoluchowski, 1959; Dexter, 1960; Howard, Vosko, and Smoluchowski, 1961). We have investigated the model by means of computer simulation, assuming a reasonable theoretical lifetime for the ionized state, and our results counteract the Varley mechanism (Chadderton, Morgan, and Torrens, 1966). Motion of the ionized particle takes place mainly along $\langle 110 \rangle$ and $\langle 111 \rangle$, but no replacement occurs, and after recombination the ion settles down in its own lattice site.

Pooley (1966) has suggested an alternative F-centre production mechanism in alkali halides which depends on the separation of the vacancy and interstitial through a $\langle 110 \rangle$ replacement sequence following radiationless electron-hole recombination. To test the feasibility of the mechanism, Pooley set up a one-dimensional computer simulation of a $\langle 110 \rangle$ line of negative ions. Neighbouring lines were represented by a periodic potential field acting on ions of the central line, and the model allowed for polarization and dielectric effects. This computation produced a value for ionic displacement threshold in a number of alkali halides which was in the region of 4 or 5 eV, which implies that permanent displacements may be produced by energies arising from electron-hole recombination. The interionic potential employed by Pooley was closely similar to that used by ourselves, suggesting that the difference in the $\langle 110 \rangle$ displacement threshold energies (25 and 5 eV) arises from another source. To check whether any part of the difference might be due to *small* differences in potential, however, we re-ran several of our computations using Pooley's Born-Mayer-Verwey interaction – with no significant alteration in the results. We therefore conclude that the differences are another consequence of the important effect of the changes in the ionic lenses. The substitution of neighbouring $\langle 110 \rangle$ lines by a periodic potential allows for the relaxation of the lenses following the passage of the traversing ion, but takes no account of the consequence of this relaxation on the interaction of the lens ions with the returning ion of the central line. Consequently, since the ionic lens does not alter in strength in the one-dimensional simulation, the return of the struck ion is hindered and it remains on a replacement site. Thus, if the ion possesses initially enough energy to pass through the potential barrier (about 5 eV), it will inevitably be permanently displaced. In the three-dimensional model the complicated interaction of the returning ion of the focusing line with the moving ions of the lens results in the return of this ion to its own site if the energy is less than about 25 eV. The full picture is evidently even more complicated since the first ion of a sequence has a greater displacement threshold than succeeding ions, both because it receives its energy instantaneously and because it costs less energy to produce *any* displacement if the process is accompanied by a replacement chain. However, we are of the opinion that the true threshold for production of a permanent displaced ion – not necessarily the primary knock-on – is well in excess of 5 eV, and on account of the ionic lens influence is likely to exceed 20 eV. Interstitial-vacancy separation at very low energies through ionic replacement sequences, with the subsequent formation of an F-centre, therefore seems unlikely. Other suggestions for colour centre production mechanisms provide for the formation of neutral atoms or molecules (Klick, 1960; Williams, 1962), and the participation of these in replacement sequences through positive hole tunnelling. These are probably more feasible than the simple ionic case, though we have no satisfactory method of testing them with the computational method.

Another consequence of the ionic potential barrier lenses has been revealed in the study of $\langle 112 \rangle$ and higher-order focusing, where a lower energy limit for focusing is evidently imposed by the asymmetry of the lens. The solid angle permitted for ions moving in this direction to generate a true focuson is small, however, and, while investigations of the $\langle 112 \rangle$ direction introduced a new type of focusing, the efficiency of the mechanism in this direction appears to be quite small. Kelly and co-workers (Kelly and Peterson, 1966) have reported results of sputtering experiments on KCl crystals and mention the $\langle 111 \rangle$ and $\langle 112 \rangle$ directions as the principal directions for ion emission from their crystals. The interpretation of these results in terms of directional effects *inside* the crystal is doubtful, since the extent to which the phenomenon of *focusing* is responsible for the appearance of preferential sputtering directions from crystal surfaces is presently in some dispute (Lehmann and Sigmund, 1966). Our computer simulations would tend to predict that $\langle 111 \rangle$ focusing has a good probability, but that $\langle 112 \rangle$ focusing is unlikely to be marked, especially for energies under 100 eV, which is the regime of energy for ions ejected in this direction during the sputtering experiment on KCl. At these low energies the ion shell repulsion collisions with the asymmetric ionic lens will cause rapid defocusing. The repulsive term in the intrinsic potential is predominant during the actual collision, and this fact takes precedence over any beneficial influence of Coulomb interaction with the asymmetric lens as a whole.

Conclusion

Within the confines of the limits imposed by the computational model the progress of fission fragment induced displacement and sub-displacement radiation damaging events in crystallites of pseudo-simple-cubic KCl and NaCl follows a similar pattern to that which has been reported for face-centred cubic copper and body-centred cubic α-iron (Gibson, Goland, Milgram, and Vineyard, 1960; Erginsoy, Vineyard, and Englert, 1964) and for hexagonal lead iodide (Chapter 10):

(1) Damage at low energies consists of vacancies and interstitials. At threshold energies the nature ·of the displacement process is strongly dependent on directional effects imposed by the lattice. At sub-threshold energies focuson propagation occurs in $\langle 100 \rangle$, $\langle 110 \rangle$, and $\langle 111 \rangle$ directions.

(2) Vacancies are of the conventional character for alkali halides but both alkali and halide ions in both salts prefer the configuration of a static crowdion along $\langle 110 \rangle$.

(3) the $\langle 110 \rangle$ direction is a favoured direction for the removal of energy from the vicinity of a primary event over a wide range of primary knock-on energies. At threshold energies the energy transfer takes place by the generation of a classical assisted focuson along $\langle 110 \rangle$. At energies in excess of 100 eV the energy transfer takes place by a repeated replacement sequence of alternate focusing and defocusing accompanied by the generation of $\langle 110 \rangle$ focusons at angles to the initial sequence.

(4) The $\langle 100 \rangle$ direction permits the generation of a simple focuson but the energy transfer process is very inefficient.

(5) The $\langle 111 \rangle$ direction is a favoured direction for strongly assisted focuson generation since it presents two triangular and symmetrical ionic lenses. At energies in excess of 100 eV, $\langle 111 \rangle$ focusons are accompanied by the generation of $\langle 110 \rangle$ focusons at angles to the initial sequence.

(6) For higher-index crystal directions such as $\langle 112 \rangle$ the asymmetry of the ionic lenses which a sequence has to traverse introduces a new kind of *impulse* focuson for which there is a lower energy limit. For energies up to 100 eV or higher, no focusing occurs and the trajectory of the initially struck ion is so modified by collision with the first strongly asymmetric ionic lens that only generation of $\langle 110 \rangle$ or $\langle 111 \rangle$ focusons at angles to the initial sequence may take place. At higher energies the trajectory of the initial struck ion may be barely modified by the collision with the lens and an impulse focuson may flow down

the high index direction. The lower limit of energy for impulse focuson generation increases with the index of the crystallographic direction.

(7) In NaCl the small size of the cation can in certain circumstances allow the simultaneous generation of channelons and focusons.

(8) The displacement threshold has a minimum along or near to the $\langle 110 \rangle$ direction for both salts. The threshold is of the order of 25 to 30 eV for KCl (potassium and chlorine) and varies between 20 eV (sodium) and 90 eV (chlorine) for NaCl. In all cases a dynamic crowdion action produces interstitials (static crowdions) at some distance from the site of a primary event, and a vacancy close to that site.

(9) The displacement threshold has an unknown maximum in excess of 150 eV along or near to the $\langle 100 \rangle$ direction. No dynamic crowdion action is observed and permanent displacement of the primary knock-on itself or an immediate neighbour is clearly favoured. Displacement along the $\langle 111 \rangle$ direction follows a similar pattern.

(10) The phenomenon of displacement and replacement in the all-important $\langle 110 \rangle$ direction is very strongly dependent on the *dynamic* behaviour of the disturbed $\langle 110 \rangle$ ionic lenses. This could account for the failure of arguments based on *static* considerations to explain the formation of colour centres after ionizing radiation.

(11) Agitations following damage events bear some resemblance to thermal spikes as they are conventionally understood, but focuson and dynamic crowdion generation along selective low-index crystallographic directions imposes a detailed structure on the cubic symmetry of the spike.

(12) A simple model for the effect of lattice vibrations indicates little effect on the progress of a typical damaging event or on the final defect configuration, though focusons are attenuated more strongly.

(13) The long-range Coulomb part of the potential is not of such importance for interactions during a radiation damage event between ions which are separated by two or more lattice parameters as might have been thought. The course of damage events is not substantially modified by variations in the magnitude of this part of the total potential, and Coulombic interaction only becomes important during relaxation, when the crystal is settling down and selecting defect sites.

References

BORN, M., and GOOEPPART-MAYER, M. (1933), *Handbuch der Physik* **24**, 633.

CHADDERTON, L. T., MORGAN, D. V., and TORRENS, I. MCC. (1966), *Phys. Letters* **20**, 329.

DEXTER, D. L. (1960), *Phys. Rev.* **118**, 934.

ERGINSOY, C., VINEYARD, G. H., and ENGLERT, A. (1964), *Phys. Rev.* **133**, A595.

FUMI, F. G., and TOSI, M. P. (1964), *J. Phys. Chem. Solids* **25**, 31.

GIBSON, J. B., GOLAND, A. N., MILGRAM, M., and VINEYARD, G. H. (1960), *Phys. Rev.* **120**, 1229.

HOWARD, R. E., and SMOLUCHOWSKI, R. (1959), *Phys. Rev.* **116**, 314.

HOWARD, R. E., VOSKO, S., and SMOLUCHOWSKI, R. (1961), *Phys. Rev.* **122**, 1406.

KELLY, J. C., and PETERSON, M. C. E. (1966), *Phys. Letters* **22**, 295.

KLICK, C. C. (1960) *Phys. Rev.* **120**, 760.

LEHMANN, C., and SIGMUND, P. (1966), *Phys. Stat. Sol.* **16**, 507.

PAULING, L. (1939), *The Nature of the Chemical Bond*, Cornell University Press.

POOLEY, D. (1966), *Proc. Phys. Soc.* **87**, 245.

ROBINSON, M. T., and OEN, O. S. (1963), *Phys. Rev.* **132**, 2385.

SZIGETI, B. (1949), *Trans. Faraday Soc.* **45**, 155.

TORRENS, I. MCC., and CHADDERTON, L. T. (1967), *Phys. Rev.* **159**, 671.

VARLEY, J. H. O. (1954), *Nature* **174**, 886.

VERWEY, E. J. W. (1946), *Rec. Trav. Chim. Pays-Bas.* **65**, 521.

WILLIAMS, F. E. (1962), *Phys. Rev.* **120**, 760.

CHAPTER TWELVE

Creation of Defects by Electron Excitation

The evidence which we have presented in the foregoing suggests that the fission fragment damage which may be observed by electron microscopy falls into two broad categories. In certain crystals, such as mica, molybdenite, alumina, etc., the damage is characterized by a linear array of defects along the trajectory of the particle (Figure 29, Chapter 5), forming a track which can be imaged by electron diffraction contrast (Hirsch *et al.*, 1960; Howie and Whelan, 1961). In other materials, however, such as metals, alloys, and certain semiconductors, the damage is much more homogeneous and consists of random arrangements of defect clusters and small dislocation loops (Figure 46, Chapter 7). We shall now show that these differences in behaviour under fission particle irradiation reflect differences in the physical properties of the materials themselves, and that the same conclusion may be reached from two quite independent starting points.

In general we know that it is possible to write a comprehensive equation for the rate of energy loss of each fission particle, neglecting nuclear transformations, as follows:

$$\left(-\frac{dE}{dx}\right)_{total} = \left(-\frac{dE}{dx}\right)_{radiation} + \left(-\frac{dE}{dx}\right)_{nuclear} + \left(-\frac{dE}{dx}\right)_{electronic} \quad (12.1)$$

The absolute stopping power is a linear sum of three terms, each of which can be considered separately. The rate of energy loss of the fission particle to radiation, to bremsstrahlung, is in fact negligible (Chadderton, Morgan, Torrens, and Van Vliet, 1966; Ch. 2) though we include it here since a very small amount of energy is dissipated by this means. The rate of energy loss in nuclear collision events, however, is certainly not negligible. Various estimates have been made of the contribution which these processes make to the arresting of a fission particle before it is deposited as an impurity atom, and they range from less than 1 % up to 10 %, according to the model which is chosen for the inter-action potential. Such treatments are based on the assumption that collision events between moving ions like fission fragments and lattice atoms at rest are describable at high energies in terms of Rutherford scattering dynamics, which favours low energy transfers, and at low energies in terms of hard sphere scattering (Chapter 2). The importance of this term in (12.1) is therefore greatest at the lower values of the energy E; and, at the higher energies, the third term, which accounts for energy given to electrons in the crystal, dominates. We do not, of course, mean to imply that the energy which is communicated to struck atoms in the crystal is not stored in lattice defects, and we do indeed suggest that, in all crystals, displacement and plasticity spikes (Brinkman, 1962), or modern modifications of these concepts such as the depleted zone (Seeger and Essmann, 1962), must lead to the introduction of Frenkel pairs. This is especially the case for metals, as the electron microscopy of gold (Chapter 7) and the alloy copper/gold has shown. It remains true, however, that, in all crystals, over 90% of the energy of each fission particle is dissipated in the energization of electrons, or delta particles, and we need to know how efficiently and how quickly these excited electrons, these waves of delta rays, communicate their energy to the lattice.

The Delta Rays

We assume that the problem of electron excitation by the fission fragment can be dealt with classically, so that the approach is relatively simple and unfettered by quantum mechanical complexities. Likewise we assume that the electrons accept the energy in a non-cooperative fashion, ignoring the possibility of plasmon emission (Pines, 1955). Now Bohr (1948) develops a comprehensive treatment of the electron excitation problem for the general case, in which he likens the atom in the crystal to an ensemble of harmonic oscillators, each consisting of an electron bound in a quasi-elastic field of force of extent:

$$a_i \sim u_i/\omega_i \qquad (12.2)$$

where ω_i is the cyclic frequency of the ith electron and u_i is an orbital velocity. By differentiating between *dynamic* 'free' collisions of short duration and 'resonance' collisions of an adiabatic character, where at any moment the atom can be essentially regarded as exposed to a *static* electrostatic field, Bohr obtains the following for the rate of energy loss:

$$\left(-\frac{dE}{dx}\right) = \frac{2\pi Z_{eff}^2 e^4 n_0}{mv^2} \sum_i \log_e \left\{ \left[\frac{\kappa}{\eta_i}\right] [\kappa]^{-2} \right\} \qquad (12.3)$$

where: $\kappa = 2Z_{eff}^2 e^2/\hbar v$, $\eta_i = 2v/u_i$; v is the velocity of the fragment; Z_{eff} is its effective atomic number; n_0 is the spatial density of lattice atoms; and m is the electronic mass. The square bracketing indicates that the quantity inside should be replaced by unity if less than one, and the summation extends over the various electrons in the atom which yield a positive, and therefore sensible, contribution to the logarithm. On a similar basis the density of excited electrons (delta particles) per unit length of fragment path in the crystal becomes:

$$W = \frac{2\pi Z_{eff}^2 e^4 n_0}{mv^2} \sum_i \frac{1}{I_i} \left\{ \left[\frac{\kappa}{\eta_i}\right]^{-1} + 2\delta_s \log_e (\eta_i [\kappa^{-1}]) \right\} \qquad (12.4)$$

where κ is a numerical factor, and $I_i = \hbar\omega_i$. Now for fission fragments over the greater part of their range $\kappa \sim 8$, and the contribution to equations 12.3 and 12.4 from 'resonance' collisions[†] is of the order of 15% only, so that in this qualitative assessment of the problem it will be convenient to employ a completely classical approach using Rutherford collisions. Then we may write:

$$d\sigma(E, T) = \frac{\pi Z_{\text{eff}}^2 e^4}{E}\left(\frac{M_1}{m}\right)\frac{dT}{T^2} \tag{12.5}$$

where T is the energy transferred to the electron and $d\sigma(E, T)$ is the differential cross-section, so that our energy-loss equation now becomes:

$$\left(-\frac{dE}{dx}\right) = n_0 \int_{T_{\min}}^{T_{\max}} T\frac{d\sigma}{dT}(E, T)\,dT \tag{12.6}$$

Summing over the i electrons in each atom and using in each case the 'ionization potential' I_i for the particular electronic shell as the lower limit of integration we obtain:

$$\left(-\frac{dE}{dx}\right) = \frac{2\pi Z_{\text{eff}}^2 e^4 n_0}{mv^2}\sum_i \log_e\left(\frac{2mv^2}{I_i}\right) \tag{12.7}$$

and for the density of free excited electrons per unit length of fragment path:

$$W = \frac{2\pi Z_{\text{eff}}^2 e^4 n_0}{mv^2}\sum_i \left(\frac{1}{I_i} - \frac{1}{T_m}\right) \tag{12.8}$$

where T_m, the maximum energy which can be transferred to an electron, is given by:

$$T_m = \frac{4M_1 m}{(M_1 + m)^2}E \simeq \frac{4m}{M_1}E \tag{12.9}$$

The *average* energy $\bar{T}$ of a typical delta particle in this simple model is given by dividing the total energy expended in ionization over an increment of fragment path by the total number of ions produced. Consequently, for a many-electron atom:

$$\bar{T} = \sum_i \left\{\log\left(\frac{T_m}{I_i}\right)\bigg/\left(\frac{1}{I_i} - \frac{1}{T_m}\right)\right\} - \bar{I} \tag{12.10}$$

and we have subtracted a term to account for the energy consumed when the electron is torn from its parent atom as the fragment passes by. Thus, in the case of irradiation of an aluminium single crystal with fission particles, a

[†] Bloch (1933), by approximating the perturbation of the wave-functions of the atomic electrons due to the incident particle, arrives at a formula which contains the quantum mechanical formula of Bethe (1930), based on the Born approximation, and the classical mechanical formula of Bohr, as limiting cases of small and large values of the parameter κ (Chapter 2).

typical delta ray carries off some 47 eV of energy, though a total of some 103 eV is expended in this 'average' ionization event. Ozeroff (1949) gives a somewhat different treatment for the case of uranium, but it is not clear what procedure he uses in order to average over the i atomic electrons for all values of the impact parameter p. Nevertheless he arrives at a figure of $\bar{T} \sim 90$ eV, which agrees very favourably with the value obtained from equation 12.10, and is greater for a heavier material, which is to be expected. In the following we shall assume that a rough but reasonable value for the energy of a typical primary electron in any fission fragment irradiated solid, light or heavy, is 80 eV. We ignore fluctuations of fragment charge and focus attention on a point in the crystal near the beginning of the fragment trajectory.

The Thermal Spike
In this, the most simple approximation, all of the energy which is liberated to electrons by the fission fragment is assumed to appear in the lattice in a very

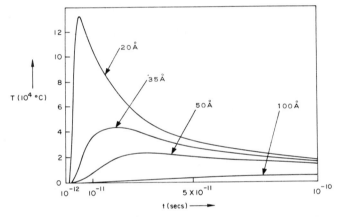

Figure 92. Temperature/time profiles for points lying at different distances from the axis of a cylindrical thermal spike produced by a typical fission fragment travelling in a crystal of lead iodide. A temperature pulse travels away from the particle, decreasing in intensity as it does so. (Curves computed from equation 12.15 with $Q = 3000$ eV/Å.)

short time, and subsequently to diffuse according to the macroscopic laws of heat conduction. The temperature distribution with time t and distance r from the axis of the spike is as follows:

$$T(r,t) = \frac{Q}{4\pi c\rho Dt} \exp\left(-r^2/4Dt\right) \tag{12.11}$$

where K is the thermal conductivity, c is the heat capacity, ρ is the mass density (diffusivity $D = K/c\rho$), and Q is the energy delivered per unit length of the spike. A temperature pulse is initiated in the lattice by the fragment and this travels out and away from the trajectory of the particle, decreasing in intensity as it does so (Figure 92).

There is no doubt that this very simple treatment does not do justice to the problem. It has been shown, however, that in certain heat-sensitive crystals the energy which is carried off by the delta rays is very rapidly communicated to the lattice, so that there is certainly no reason to discard the thermal spike as being of no further significance (Fleischer, Price, and Walker, 1965). Moreover, the computer simulation studies, made by Vineyard (1961) and his associates, of the dynamic motions of atoms in an irradiated lattice reveal that, after 10^{-12} or 10^{-11} sec, the concept of a spherical thermal spike (Seitz and Koehler, 1956) can be shown to have very real validity.

It is usual to avoid the difficulty of a singularity at zero time by considering only the course of events from a starting time:

$$t_0 = r_s^2/4D \qquad (12.12)$$

where r_s, the mean atomic radius (Seitz and Koehler, 1956), is defined by the equation:

$$\tfrac{4}{3}\pi r_s^3 = 1/n_0 \qquad (12.13)$$

For the fission thermal spike (equation 12.11), therefore, and for a typical solid with a melting temperature T_m, the distance r_m from the axis to which the high temperature boundary has diffused when the temperature has dropped to T_m is determined by:

$$\pi r_m{}^2 n_0 = Q/3kT_m \qquad (12.14)$$

Hence, substituting for n_0 from (12.13):

$$r_m = \tfrac{2}{3}(Q/kT_m)^{1/2} r_s^{3/2} \qquad (12.15)$$

Furthermore, if we neglect the exponential part of equation 12.11 and consider a disordering process with an activation energy E', then the probability, or the number of times n_j for which the process may be expected to occur in the central zone of the spike during the period of duration, per unit length of spike, is given by:

$$n_j = \int_0^\infty N\,dt = \int_0^\infty N_0 \exp(-E'/kT(t))\,dt \qquad (12.16)$$

which may be shown to become

$$r_j = \Lambda r_s N_a/9 \qquad (12.17)$$

where $\Lambda = \nu_0 r_s^2/D$, and $N_a = Q/E'$ is an 'activation number'. Likewise the critical distance r' within which the probability that a given atom participates in the disordering process is unity becomes:

$$r' = \tfrac{2}{3} r_s^{3/2}[N_a \log_e r_j]^{1/2} \qquad (12.18)$$

and the number of atoms which actually participate per unit length of spike, some of them more than once:

$$n_{\mathrm{a}} = \frac{n_{\mathrm{j}} N_{\mathrm{a}}}{3} \exp \left\{ -\frac{9r'^2}{4 N_{\mathrm{a}} r_{\mathrm{s}}^3} \right\} + \frac{3r'^2}{4r_{\mathrm{s}}^3} \qquad (12.19)$$

The variation of r' and n_{a} as a function of a modified activation number N'_{a} (Q'/E', where Q' is the energy loss per mean atomic radius) is shown in Figure 93 for the case of uranium ($E' = 3\,\mathrm{eV}, \Lambda = 1$).

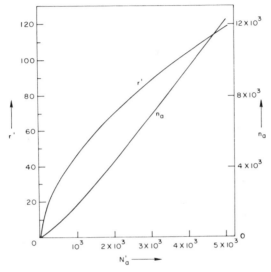

Figure 93. Variation of r' and n_{a} as a function of the modified activation number N'_{a} for a fission fragment travelling in uranium. (See text.)

The Classical Energy Transfer between Delta Particles and Lattice

The simple treatment of the previous section clearly is insufficient since it omits any consideration of the fundamental differences in the electron/lattice energy transfer processes for the different crystals. When the primary electrons, or delta particles, are energized by the fission fragment, and are removed from their parent atoms, they begin to follow a tortuous path through the crystal and undergo multiple scattering collisions. One way to deal with this kind of problem (Brooks, 1950) is to establish a simple classical model for the state of affairs which the fission fragment leaves in its wake. It is assumed that a Maxwellian gas of electrons (temperature T_1) is homogeneously mixed with a second Maxwellian gas of heavy atoms (temperature T_2), and that the rate of energy interchange between the two systems is very much less than that within each system, so that each system is in equilibrium. Instead of equation 12.11

we may then write the following for the temperature of the lattice (see Appendix 2):

$$T_2 = \frac{Q_1}{2\pi C_1} \exp(-\mu t) \int_0^\infty \exp(-(D_1 + D_2)k^2 t/2) \frac{\sinh(\mu t\sqrt{(1 + \delta^2)})}{\sqrt{(1 + \delta^2)}} J_0(kr)k\, dk$$

(12.20)

where $\delta = (D_1 - D_2)k^2/2\mu$ and D_1 and D_2 are diffusivities determined by the thermal conductivities (K_1, K_2) and heat capacities per unit volume $(C_1 = C_2 = C)$ of the electrons and lattice respectively $(D_1 = K_1/C_1, D_2 = K_2/C_2)$, and where $\mu = b/C$ allows for the coupling between the two systems. In this equation we can vary the relative magnitudes of D_1 and D_2, and also the absolute magnitude of the coupling parameter μ, so that it is possible to simulate conditions for the principal crystal types.

Metals

The case of uranium has been considered in detail by Brooks (1950) and by Ozeroff (1949). Equation 12.20, however, can be applied quite generally to the metals if we make the reasonable assumption that $D_1 \gg D_2$ $(K_1 \gg K_2)$, when for $r = 0$ it becomes:

$$T_2 = \frac{Q_1}{4\pi C} \frac{\mu \exp(-\rho)}{(D_1 - D_2)} \sum_{n=0}^\infty \Gamma(n + \tfrac{1}{2})\left(\frac{2}{\rho}\right)^{(n+1/2)} \left(\frac{1}{g}\right)^{2n} I_{(n+1/2)}(\rho) \quad (12.21)$$

where $\rho = \mu t$, $g = (D_1 + D_2)/(D_1 - D_2)$, and $I(\rho)$ is a Bessel function of imaginary argument. By rearrangement of terms, equation 12.21 may be expressed in an ascending power series:

$$T_2 = \frac{Q_1}{2\pi C} \frac{\mu \exp(-\rho)}{(D_1 - D_2)} \sum_{m=0}^\infty \frac{1}{m!}\left(\frac{\rho}{2}\right)^{2m} \sum_{n=0}^\infty \left\{\frac{\tfrac{1}{2}g^{-2n}}{(n + \tfrac{1}{2})\dots(n + m + \tfrac{1}{2})}\right\} \quad (12.22)$$

Now, if the contribution of heat to the lattice from the electrons were neglected, only the contribution to the thermal spike from nuclear collisions would remain $((-dE/dx)_{nuclear}$ in equation 12.1). Let us suppose that on average these processes lead to the deposition of Q_2 eV per unit length of spike, and that the rate of energy deposition in the electron system $((-dE/dx)_{electronic}$ in equation 12.1) $Q_1 \sim 20Q_2$. If only nuclear collisions were accounted for, therefore, the temperature/time distribution at the origin would be given by:

$$T_2^* = \frac{Q_2}{4\pi C} \frac{1}{D_2 t} \quad (12.23)$$

Hence, writing $g \sim 1 + \epsilon$, where $\epsilon = 2D_2/(D_1 - D_2)$ and is small:

$$\frac{T_2}{T_2^*} = 2 \frac{Q_1}{Q_2}\left\{\frac{D_2}{D_1 - D_2}\right\}\rho \exp(-\rho) f(\rho, \epsilon) \quad (12.24)$$

where $f(\rho, \epsilon)$ represents the series. Then, for the case of uranium ($\epsilon \sim 1/300$), evaluating the first four terms in the power series to terms of order 6:

$$\frac{T_2}{T_2^*} = 0\cdot204\, \rho \exp(-\rho)\{1 + (7\cdot6)10^{-2}\rho^2 + (3\cdot25)10^{-3}\rho^4 + (4\cdot5)10^{-5}\rho^6 \ldots\} \quad (12.25)$$

This expression has a broad maximum of $0\cdot08$ at $\rho \sim 1\cdot2$. In other words, the maximum temperature in a metal lattice which is communicated to it by the excited electrons is *less than 10%* of that due to the heat directly communicated by nuclear collisions, a conclusion which is independent of the relaxation parameter μ.

For finite values of the distance r from the spike axis the evaluation is most easily determined by integrating equation 12.20 over a normalized Gaussian volume region.

Semiconductors

For the situation in semiconductors we assume that $D_1 = D_2$, so that $\delta = 0$. The solution to the integral in (12.20) is then trivial and the temperature in the lattice becomes:

$$T_2 = \frac{Q_1}{2C} \cdot \frac{1}{4\pi D_2 t} \exp(-r^2/4D_2 t)\{1 - \exp(-2\mu t)\} \quad (12.26)$$

The ratio of the two lattice temperatures, however, is given by:

$$\frac{T_2}{T_2^*} = \frac{Q_1}{2Q_2}(1 - \exp(-2\mu t)) \quad (12.27)$$

Assuming once again that $Q_1 \sim 20Q_2$, this indicates that the temperature in a semiconductor lattice at *any* point will be $10(1 - \exp(-2\mu t))$ time *higher* than the temperature in the lattice due to nuclear collisions only.

Insulators

For the case of insulators we may write that $D_2 \gg D_1$ since the lattice conductivity is dominant. Now, however, $t = 2D_1/(D_2 - D_1)$ and is small, and g becomes $-(1 + \epsilon)$, so that while we once again obtain equation 12.24 the series $f(\rho, \epsilon)$ is now somewhat different. In fact if the new ϵ is $\sim 1/300$:

$$\frac{T_2}{T_2^*} = 128\, \rho \exp(-\rho)\{1 + (7\cdot65)10^{-2}\rho^2 + (3\cdot13)10^{-3}\rho^4 + (7\cdot25)10^{-4}\rho^6 + \ldots\}$$
$$(12.28)$$

Thus the maximum temperature communicated to a 'typical' insulating lattice by the excited electron system is some *fifty times higher* than that communicated directly.

This simple classical mixed-gas model therefore suggests that after the passage of a fission fragment the lattice thermal spike is most intense in

14

insulators, least intense in metals, and somewhat intermediate in semi-conductors (Figure 94). Hence if the profile of the spike is broad and flat in metals the 'effective' rate of energy delivery Q of our earlier model (equation 12.11) must be correspondingly small. Similarly the activation number N_a must also be small since it is directly proportional to Q, and this implies that very few lattice atoms (n_a in equation 12.19) will participate in activated dis-ordering processes, such as neighbour atom interchange or point defect creation and migration. In insulators, on the other hand, since the electronic energy is rapidly communicated to the lattice, the temperature pulse will be sharply peaked at the axis, and if E' is not too small many defects will result.

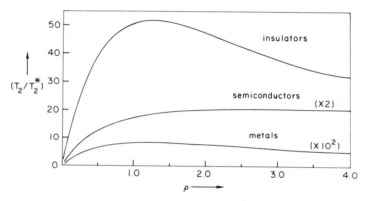

Figure 94. Variation of the lattice temperature ratio T_2/T_2^* as a function of the coupling parameter ρ for insulators, semiconductors, and metals.

The Quantum Mechanical Energy Transfer between Delta Particles and Lattice

While the crude classical mechanical analysis which we have so far established gives us some kind of feeling for the importance of the fundamental electron/atom scattering process it can hardly be described as satisfactory. In the first instance it is not necessarily true that the rate of energy interchange between the two systems – electron and atom – is very much less than that within each system. Secondly the atoms are not free like the atoms of a Maxwellian gas but are bound to sites on a three-dimensional space lattice, so that they recoil 'in unison' when struck by an energetic electron. We should really concern ourselves with phonons, with quantized vibrations of the lattice, which are a fundamental property of the nature of the irradiated material.

We can see this in a qualitative way if we make a crude comparison between the different crystal types. Consider first of all a metal. The existence of a 'sea' of free electrons, and of an infinite system of energy levels above the Fermi surface, implies that electron/electron collisions will be of prime importance. The primary electron therefore undergoes high-angle scattering collisions in producing secondaries and follows a devious route through the crystal until it joins the conduction electrons. Before being degraded to this level, however, the primary, and its secondary progeny, must communicate energy to the lattice in electron/phonon collisions. In insulators it is probably permissible to neglect the very small density of electrons in the conduction band. The energy losses of the primary in falling initially from 80 eV must therefore be in excitation of secondary electrons across the forbidden gap between the valence and conduction bands. This gap will be of the order of several electron volts so that, just as soon as the primary energy is less than this value, electron/phonon interactions will prevail. The intrinsic semiconductor will undoubtedly be something of an intermediate case. Since the upper filled band is separated from the conduction band by only about 1 eV the primary electrons, in falling to this value, will lose appreciable amounts of energy by exciting lattice electrons. Thereafter phonon emission will become important. What we are concerned with, therefore, are the relative magnitudes of the electron/electron and electron/phonon energy loss rates in the different crystal types and as the energy of the primary electron falls. We shall treat the problem simply, using approaches which have been useful in explaining the processes of secondary electron emission and dielectric breakdown.

Metals

The metals have been considered by Seitz (1944) in wartime work on the Manhattan project and more recently and in considerable detail by Goland and Paskin (1964). Moreover, since the electrons in a metal are free, we have a situation which closely corresponds to the classical situation which we discussed previously.

Suppose that two electrons in a metal single crystal approach each other at a distance corresponding to the impact parameter p. The angle γ through which they are deflected in the centre of mass system is then given by the relation:

$$\tan\left(\frac{\gamma}{2}\right) = \frac{e^2}{2E_p} \tag{12.29}$$

where E is the energy of the moving electron in the original frame of reference. Since the energy the moving electron transfers to that with which it collides is $E\sin^2(\gamma/2)$, it follows that the energy transferred, ΔE, is related to p as follows:

$$\Delta E = E\frac{a^2/p^2}{(1 + a^2/p^2)} \tag{12.30}$$

where $a = e^2/2E$. The energy the moving electron loses in travelling unit distance to the electrons which lie in a cylindrical shell which is centred about the path of the electron and whose walls are bounded by circles of radii p and $p + \mathrm{d}p$ is:

$$E\frac{a^2/p^2}{(1 + a^2/p^2)} 2\pi n_0^{(e)} p\,\mathrm{d}p \tag{12.31}$$

in which $n_0^{(e)}$ is the density of electrons. Thus the rate at which the moving electron gives up its energy to its neighbours is given approximately by:

$$\left(-\frac{dE}{dx}\right) = \frac{\pi e^4 n_0^{(e)}}{2E} \log_e \left\{\frac{2p_0 E}{e^2}\right\} \qquad (12.32)$$

where p_0 is the value of the impact parameter at which the moving electron is screened from its neighbours. In passing from (12.31) to (12.32), the numerator in (12.31) has been replaced by unity and the lower limit of integration has been adjusted in such a way as to avoid a singularity that would result from this simplification.

Now p_0 is of the order of magnitude of the interatomic distance so that the logarithm in (12.32) is near to unity. Furthermore, we can also conclude from the magnitude of the coefficient of the logarithm that in moving one interatomic distance the electron loses an energy which is comparable with its initial energy, which implies that it will rapidly come into temperature equilibrium with its neighbours.

The rate at which the electrons transfer their energy to the lattice is comparatively small, however. It is readily shown that if a moving electron and a stationary lattice atom make a collision in which the particles are deflected through an angle γ in the centre of mass co-ordinate system, the energy lost by the moving electron is approximately given by:

$$\Delta E = 4E\frac{m}{M}\sin^2\left(\frac{\gamma}{2}\right) \qquad (12.33)$$

where m is the electronic mass and M the mass of the lattice atom. Comparing this equation with (12.30) and (12.32) we can conclude that the rate at which the electron loses energy to the lattice is given approximately by the equation:

$$\left(-\frac{dE}{dx}\right) = \frac{2\pi e^4 n_0}{E}\frac{m}{M}\log_e\left\{\frac{2pE}{e^2}\right\} \qquad (12.34)$$

and the coefficient in this relation is about 10^5 times smaller than the corresponding coefficient in (12.32).

These general conclusions are borne out by the Goland and Paskin (1964) treatment. The ratio of the energy loss rates for the two processes:

$$\left(-\frac{dE}{dt}\right)_{ee} \Big/ \left(-\frac{dE}{dt}\right)_{ep} = \left(\frac{E}{kT}\right)\left(\frac{\tau_{ep}}{\tau_{ee}}\right) \gg 1 \qquad (12.35)$$

evidently indicates that the electron/electron interaction is a much more rapid mechanism for energy transfer because its relaxation time τ_{ee} is initially shorter than that of the electron/phonon process and because it transfers more energy per collision (i.e. $E/kT \gg 1$). Down to energies less than 1 eV the electron/electron interaction dominates. Moreover, as we shall see in discussing semiconductors, when the electron energies approach the Fermi level the collision probability for interaction with the acoustic mode phonons falls, and the total effective range x_p of the primary electron is therefore large (Quinn, 1962). As a direct consequence of these several effects it becomes clear that by the time the electron energy has been passed to the lattice itself it is distributed over a large volume (Seitz and Koehler, 1956), and that Whiddington's law for the rate of energy loss of the primary electrons:

$$(-dE/dx) = -A/E \qquad (12.36)$$

is probably a good approximation to (12.32) and to a more accurate formula of the Bethe/Bloch kind (Dekker, 1958). Dekker and Van der Ziel (1952) and Van der Ziel (1953) have treated the same problem quantum mechanically on the assumption of both simple and screened Coulomb interactions, and their conclusions bear out the results of comprehensive classical treatments presented earlier by Kadyschewitsch (1940) and Baroody (1950).

Semiconductors

For both the intrinsic semiconductors and the insulators we shall adopt a simple model in which it is assumed that a Whiddington law will apply down to primary kinetic energies with an order of magnitude of the fundamental energy band gap E_g. In the semiconductors this implies that the law applies down to kinetic energies in the region of 5 eV. *All* electrons with a kinetic energy E_s which is less than this, and which are unable to free further electrons, will be defined as secondaries, and it will be assumed that *all* of their energy is subsequently passed to the lattice. We shall be interested in obtaining a total effective secondary electron range x_s since this will be a measure of the efficiency with which the lattice absorbs the electronic energy.

In the general case we should consider the excitation of optical and acoustic modes by the secondary electron, plasmon/phonon interactions, Laue scattering, and the generation of excitation waves (excitons). We shall only consider the single electron/phonon emission processes, neglecting other stopping effects on the assumption that they are negligible.

For the acoustic modes Seitz (1949) has shown, using time-dependent perturbation theory and the deformable atom model (Bloch, 1928; Ziman, 1963), that as long as the electron wave vector **k** lies well inside the first Brillouin zone the collision frequency f can be expressed as follows:

$$f = \frac{4}{9\pi} \frac{C^2 kT}{h^3 c^2 n} \frac{m}{M_2} \mathbf{k} \qquad (12.37)$$

where M_2 is the mass of a lattice atom, n is the density of unit cells, c is the velocity of the longitudinal modes, and C is an energy parameter describing the coupling between electron and lattice. A similar relation for collisions with optical modes may also be derived (Seitz, 1948) so that the ratio of the energy loss in collisions with acoustic modes of vibration to the loss in collisions with optical modes becomes:

$$R = \left(\frac{16}{9}\right)\left(\frac{C^2}{D^2}\right)\left(\frac{\mathbf{k}^2}{\mathbf{K}^2}\right) \qquad (12.38)$$

where D is once again a coupling parameter and $\mathbf{K}$ is the first nonvanishing reciprocal lattice vector. What little experimental evidence there is available on these matters suggests that D is somewhat less than C and that the influence of optical modes may be almost negligible. In either event, in both cases the collision frequencies vary from about 10^{13} sec^{-1} at energies of the order of kT to about 10^{15} sec^{-1} when the wave number vector approaches the zone boundary, where there is a maximum in the so-called 'electron friction'. For semiconductors therefore, we shall only consider the emission of acoustic modes by the secondary and ignore optical mode losses.

The rate of energy loss to acoustic modes is given by an expression which is derived in a similar manner to (12.37):

$$\left(-\frac{dE_s}{dt}\right) = \frac{4}{9\pi} \frac{C^2}{\hbar} \frac{\mathbf{k}^3}{n} \frac{m}{M_2} \tag{12.39}$$

and, knowing the velocity v, (12.37) and (12.39) can be used to obtain the rate of energy loss with distance $(-dE/dx)$ and λ, the mean-free-path for the scattering process. The energy loss in each collision then becomes:

$$\alpha_1(T, E_s) = \left(-\frac{dE_s}{dx}\right)\lambda_1 = \frac{(\hbar c \mathbf{k})^2}{kT} = \frac{2mc^2 E_s}{kT} \tag{12.40}$$

and this is temperature and energy dependent.

The model is now simplified further by assuming that we can adopt an average energy loss at each collision, independent of energy, such that:

$$\alpha_1(T) = \frac{mc^2 E_s}{kT} \tag{12.41}$$

The incremental effective range x'_s of a typical secondary of energy E_s, in falling to an energy E', can then be obtained by assuming that it executes a one-dimensional Brownian motion (Appendix 3) owing to phonon collisions; an approach which has been used by Dekker (1954) in an analysis of secondary electron emission. The expression derived:

$$x'_s = \frac{9\pi\hbar^4 c^2 n M_2}{4C^2 m^2 (kT)^{1/2}} \left\{ \frac{E_s - E'}{mc^2 E_s} \right\} = \left\{ \left(\frac{A}{mc^2}\right) \left(\frac{E_s - E'}{E_s}\right) \right\}^{1/2} \tag{12.42}$$

indicates clearly that in the non-polar semiconductors the excited electron travels a *smaller* effective distance in falling from a kinetic energy E_g to $E_g/2$, than it does in falling from a kinetic energy $E_g/2$ to $E_s = 0$; a direct consequence of the fact that collisions with acoustic modes are less probable at lower energies.

Insulators

The case of polar crystals – insulators – is much more complex. Although we can once again ignore the normal optical modes, there are, in addition to collisions with acoustic modes, collisions with polar modes (Fröhlich and Mott, 1939; Fröhlich, 1939) and neither can be neglected (Seitz, 1949). For the polar modes only, the mean-free-path between collisions may be shown to be given by:

$$\frac{1}{\lambda_2} = \frac{\pi m Z^2 e^4}{2n' M_2' a^5 h\nu} \frac{(2n_\nu + 1)}{E_s} \int \frac{dq}{q} \qquad (12.43)$$

where M_2' is the mass of an ion pair, a is the nearest neighbour distance, Ze is the effective ionic charge, ν is the frequency of the longitudinal modes, n' is the number of ion pairs, E is greater than $h\nu$, n_ν is $(\exp(h\nu/kT) - 1)^{-1}$, and where the integration $\int dq/q$ over the wave numbers q of the lattice waves is approximately a constant. The lower limit of the integration over q is equal to $2\pi m\nu/\hbar k$, where $\mathbf{k}$ is the wave vector of the electron. Likewise the rate of energy loss of the electron is:

$$\left(-\frac{dE_s}{dx}\right) = \frac{\pi m Z^2 e^4}{2n' M_2' a^6} \frac{1}{E_s} \int \frac{dq}{q} \qquad (12.44)$$

so that, on the average, the energy lost per collision becomes:

$$\alpha_2(T) = \left(-\frac{dE_s}{dx}\right)\lambda_2 = \frac{h\nu}{(2n_\nu + 1)} \qquad (12.45)$$

The incremental effective range (Appendix 3) of the secondary electron may then be expressed as follows:

$$x_s' = \frac{2n' M_2' a^6 (h\nu)^{1/2}}{\pi m Z^2 e^4} \left(\int \frac{dq}{q}\right)^{-1} \left\{\frac{(E_s - E')(E_s^2 + E'^2 + E_s E')}{3(2n_\nu + 1)}\right\}^{1/2}$$

$$= \left(\frac{B}{h\nu}\right)^{1/2} \left\{\frac{(E_s - E')(E_s^2 + E'^2 + E_s E')}{3(2n_\nu + 1)}\right\}^{1/2} \qquad (12.46)$$

which suggests that the electron would travel a *greater* effective distance in falling from a kinetic energy E_g to $E_s' = E_g/2$, than it would in falling from a kinetic energy $E_g/2$ to $E_s' = 0$; a consequence of the fact that collisions with polar modes are more probable at lower energies.

It is a difficult matter to include collisions with both acoustic *and* polar modes of vibration. We accomplish this in some measure, however, by assuming a mean-free-path for scattering which is compounded of both processes:

$$\bar{\lambda} = (\lambda_1 + \lambda_2)/2 \qquad (12.47)$$

so as to make the analysis of Appendix 3 tractable for this complicated case. In this approximation the incremental effective range becomes:

$$x_s' = \left\{\left[\frac{(E_s - E')}{(\alpha_1(T) + \alpha_2(T))}\right]\left[\frac{A}{kT} + \frac{B(E_s^2 + E'^2 + E_s E')}{3(2n_\nu + 1)^2} + \left(\frac{AB}{kT}\right)^{1/2}\frac{(E_s + E')}{(2n_\nu + 1)}\right]\right\}^{1/2}$$

$$\qquad (12.48)$$

where A and B are defined in equations 12.42 and 12.46. This gives a much smaller value for the total effective range x_s than either (12.42) or (12.46) alone.

This simplified quantum mechanical electron/phonon scattering model therefore suggests that the total effective range of a typical secondary electron, produced by typical delta particles during fission fragment irradiation of a

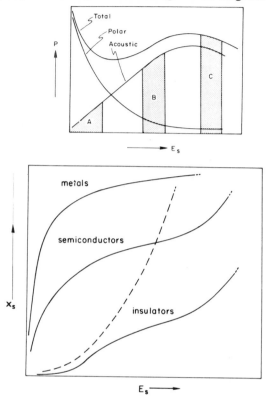

Figure 95. (a) Schematic representation of the collision frequency p as a function of electron energy E_s, in polar and non-polar crystals. In metals, where only acoustical modes are responsible for scattering, the electron is rapidly brought down to region A by collisions with free electrons. In non-polar crystals such as germanium or diamond, collisions with free electrons cease in the vicinity of the valence/conduction band gap when collisions with acoustic modes begin to dominate (region B). In polar crystals such as sodium chloride, collisions with both acoustic and polar modes become important when the electron energy falls below the band gap (region C). There are two peaks in the total 'electron friction'. (Regions A, B, and C on the energy scale have only qualitative value, and no quantitative significance is to be inferred from their position.) (b) Schematic representation of the way in which the secondary electron range x_s varies as a function of the energy E_s for insulators, semiconductors, and metals. The broken line indicates how the range would vary if only polar modes were responsible for scattering.

crystalline solid, is greatest in metals, least in polar materials, and intermediate in non-polar crystals (Figure 95). The rate at which the energy of the fission particle appears in the lattice must be largely dependent on this important parameter.

Discussion

On the basis of two independent models – one classical and one quantum mechanical – it has been shown that the electron excitation in a fission spike is of short duration in insulators, lasts longer in semiconductors, and exists for the longest time in crystals of metallic solids. In the classical approximation, these differences reflect the efficiency with which a 'lattice gas' can act as a sink for electronic energy, while in the simplified quantum mechanical model they are a direct result of the nature of the lattice vibrations which can be excited in the crystal. It should be clear, however, that both approaches are radical simplifications of the very complicated processes which must take place during and after onset of the fission spike; processes which are a sensitive function of the detailed nature of the lattice array. Bullough and Gilman (1966), for example, propose that in certain solids the rapid delivery of the energy of the fission fragment to the lattice may mean that the material is effectively 'inertially' clamped for a brief period of time. A solution of the wave equation then exists which implies that the sudden absorption of energy into a cylindrical region can lead to exceedingly high negative pressures which may in turn produce fracture or other damage. A further hypothesis and one we shall consider in detail later (Chapter 13) is that of 'ion explosion' (Fleischer *et al.*, 1965). This is essentially an extension of the ideas of the Varley mechanism (1954) to all solids. It is assumed that in the wake of the fission particle there is a dense core of positive charge on account of multiple ionization processes induced by Coulomb interactions with the passing particle, and that this has a sufficiently long lifetime that explosive Coulomb repulsion can eject ions into the surrounding lattice to produce Frenkel pairs. Serious difficulties with this proposal, however, arise out of doubts about the lifetime of the excited ions, on which the model crucially depends, and from misunderstandings concerning the dynamic influence of the atomic 'lenses' which border the particle trajectory (Chadderton, Morgan, and Torrens, 1966). Furthermore, one can never be certain what parts the processes of 'focusing' and 'channelling' play (Chadderton, 1965) in such complicated dynamic events.

In the simplified models discussed in this chapter the differences in the nature of the fission particle damage observed by electron microscopy in different solids derive substantially from the rate at which the energy appears in the lattice, though surface effects, anisotropy of the lattice, and complicated 'explosion' effects of the kind we have briefly mentioned may complicate the issue. In either event once the energy has arrived in the lattice it is assumed that the cylindrical thermal spike as it is conventionally understood has some significance, so that a calculation of the number of atoms n_a participating in activated rate processes is of value (equation 12.19). This number is itself clearly strongly dependent on the rate at which the energy is arriving in the lattice and, in the simple quantum mechanical model, for example, it will be inversely proportional to some power of the total effective secondary electron range x_s. Suppose that we consider three possible activated rate processes:

(i) Frenkel defect creation; (ii) neighbour atom interchange; (iii) point defect migration; and arbitrarily assume that for all solids the activation energy is identical for (i) and (ii), but is somewhat less for (iii). In metals like gold, n_a will be small ($\sim$several hundred per mean atomic radius) since x_s is large ($\sim$several hundred Ångström units), and (ii) will not produce disorder since we are considering an element. In ionic solids like sodium chloride, on the other hand, n_a will be large ($\sim$several thousand per mean atomic radius) since x_s is small ($\sim$several tens of Ångström units), so that more defects will be formed by (i), and, though process (ii) will probably be followed by Coulomb repulsion, permanent damage may result since we are considering a compound. Whether the defects remain near the initial trajectory of the fission particle will depend upon the part played by process (iii) and the extent to which it is favoured. In metals it is our experience that point defect migration proceeds very easily at room temperatures, so that (iii) will be encouraged, while in ionic and covalent compounds this will not necessarily be the case.

This general picture we believe to be substantiated by the majority of the experimental results. In only one case, graphite, have fission particle tracks been reported in an element (Izui and Fujita, 1961), and in spite of several attempts we have not been able to repeat this result. Moreover, in compounds it is frequently possible to ascribe an apparently anomalous behaviour to an unusually high point defect mobility, as in the case of crystals of irradiated lead iodide (Chapters 9 and 10) and this is often a manifestation of anisotropy of crystal structure. We therefore conclude that the experimental results are in general accord with the findings of both of the models presented in this chapter, namely that the ranges of the secondary electrons, or the magnitude of the electron/lattice coupling factor, are all-important as a measure of the rate at which the energy of the fission particle appears in the lattice proper.

References

BAROODY, E. M. (1950), *Phys. Rev.* **78**, 780.

BETHE, H. (1930), *Ann. Phys.* **5**, 325.

BLOCH, F. (1928), *Z. Phys.* **52**, 555.

BLOCH, F. (1933), *Ann. Phys.* **16**, 285.

BOHR, N. (1948), *Math-fys. Medd.* **18**, 8.

BOWDEN, F. P., and CHADDERTON, L. T. (1962), *Proc. Roy. Soc. A* **269**, 143.

BOHR, N. (1948), *Math-fys. Medd.* **18**, 8.

BOWDEN, F. P., and CHADDERTON, L. T. (1962), *Proc. Roy. Soc. A* **269**, 143.

BRINKMAN, J. A. (1962), N. Amer. Aviation Rep., SR-6642.

BROOKS, H. (1950), AEC Research Report No. TID-10033.

BULLOUGH, R., and GILMAN, J. J. (1966), private communication.

CHADDERTON, L. T. (1964), *Proc. Roy. Soc. A* **280**, 110.

CHADDERTON, L. T. (1965), *Radiation Damage in Crystals*, Methuen (London).

CHADDERTON, L. T., MORGAN, D. V., and TORRENS, I. MCC. (1966), *Solid State Comm.* **4**, 391.

CHADDERTON, L. T., MORGAN, D. V., TORRENS, I. MCC., and VAN VLIET, D. (1966), *Phil. Mag.* **13**, 185.

DEKKER, A. J. (1954), *Phys. Rev.* **94**, 1179.

DEKKER, A. J. (1958), *Solid State Physics* **6**, 251.

DEKKER, A. J., and VAN DER ZIEL, A. (1952), *Phys. Rev.* **86**, 755.

FLEISCHER, R. L., PRICE, P. B., and WALKER, R. M. (1965), *J. Appl. Phys.* **36**, 3645.

FRÖHLICH, H., and MOTT, N. F. (1939), *Proc. Roy. Soc.* **171**, 496.

FRÖHLICH, H. (1939), *Proc. Roy. Soc.* **172**, 94.

GOLAND, A. N., and PASKIN, A. (1964), *J. Appl. Phys.* **35**, 2188.

HIRSCH, P. B., HOWIE, A., and WHELAN, M. J. (1960), *Phil. Trans. A* **252**, 499.

HOWIE, A., and WHELAN, M. J. (1961), *Proc. Roy. Soc. A* **263**, 217.

IZUI, K., and FUJITA, J. E. (1961), *J. Phys. Soc. Japan* **16**, 1032.

KADYSCHEWITSCH, A. E. (1940), *J. Phys. USSR* **2**, 115.

OZEROFF, J. (1949), AEC Research Report No. AECD-2973.

PINES, D. (1955), *Solid State Physics* **1**, 367.

QUINN, J. J. (1962), *Phys. Rev.* **126**, 1452.

SEEGER, A., and ESSMANN, U. (1962), Enrico Fermi Summer School on Radiation Damage, Ispra, Italy.

SEITZ, F. (1948), *Phys. Rev.* **73**, 550.

SEITZ, F. (1949), *Phys. Rev.* **76**, 1376.

SEITZ, F. (1944), AEC Research Report No. CP-1598.

SEITZ, F., and KOEHLER, J. S. (1956), *Solid State Physics* **2**, 305.

VAN DER ZIEL, A. (1953), *Phys. Rev.* **92**, 35.

VARLEY, J. H. O. (1954), *Nature* **174**, 886.

VINEYARD, G. (1961), *Energia Nucleare* **8**, 9.

ZIMAN, J. M. (1963), *Electrons and Phonons*, Oxford University Press.

CHAPTER THIRTEEN

Spikes of Electronic Excitation

In a detailed analysis of the creation of defects by electronic excitation we have been considering the traditional approach to the problem of the 'thermal spike'. Kinetic energy received by the electrons is passed to the lattice at a rate which depends upon the fundamental properties of the material, and appears as heat. Thus the concept of a 'spike' in the terminology of radiation damage physics is intimately associated with the release of energy. When delta particles are energized in a solid by a passing fission fragment, however, not all of the energy which the fragment loses appears immediately in the kinetic form. Some is consumed in overcoming the binding energy of the electrons to their respective nuclei and is stored for a somewhat longer period as electrostatic potential energy. It is just as important for us to consider the fate of this energy and to assess its own ultimate effectiveness in the creation of lattice defects. Is the energy permanently stored, or is there a rapid and efficient conversion to more damaging heat?

The Ion Explosion Spike

This new and different type of spike, which provides for the formation of an interstitial-rich shell around a vacancy core from which electrons have been removed by ionization, is introduced to make allowance for the effect of electrostatic repulsive forces in that ionized region of the solid which marks the 'track'. The model envisaged is that of a narrow cylinder of ionized atoms surrounding the charged particle path. Since this contains an excess of positive ions which repel each other and distort the lattice, it might be expected that some would be ejected into interstitial positions leaving vacancies in a central core, surrounded by a cylindrical region of interstitials. If the lattice subsequently collapses around this vacancy core a charged particle track may be formed, delineated by a trajectory of defects. The true extent of the damage formed will in fact depend critically on the lifetime of the ionized atoms before electron capture. An associated problem in the Varley mechanism (Varley, 1954) is still unanswered, and, since the present question is a complication of that basic problem, no reliable estimates of the lifetime are available. However, a semiquantitative treatment (Fleischer, Price, and Walker, 1965) concludes that, in the case of insulators with low hole mobility, the appropriate lifetime should be in the region of 10^{-12} to 10^{-13} sec.

We may approach the problem from another viewpoint – that of the dynamics of the displacement process – and try to obtain answers to three fundamental questions. Firstly, what type of lattice distortion may we expect following such a multiple ionization process? Secondly, how long will it take for any damage to form? Thirdly, will any of this damage be stable after electron capture? The dynamics of this problem may be conveniently simulated for a KCl structure on a high-speed digital computer by a procedure similar to that which was described in Chapter 11. It may be assumed that a charged particle travelled in the $\langle 100 \rangle$ direction of KCl, ionizing consecutive $\{100\}$ planes continuously in its path, and the computer can follow the dynamic atomic motions in one of these planes, the $\{100\}$ plane being the most densely populated, thus giving the maximum effect of the spike. The two-dimensional approximation is acceptable in this case for the following reasons:

(1) If the ionization is uniform and identical for successive planes, then the forces on an ion from one side, which try to force it out of the plane, will on average be counteracted by equal and opposite forces from the other plane.
(2) The major unbalanced force on any ion will be the strong outward electrostatic force *in* the plane owing to the ionized atomic core.

A planar crystallite of KCl (20×20 ions without any lattice vibrations) was accordingly simulated in the computer (Chadderton, Morgan, and Torrens, 1966). Four chlorine ions near the centre of the array were multiply ionized, and the subsequent motions of ions in the array calculated. The interaction potential used was identical to that employed for KCl in Chapter 11, with the exception that the *electrostatic* term was multiplied by a parameter N, characteristic of the ionic interaction which was varied depending on the interacting pair of ions:

$N = 1$ two normal lattice ions (K^+–K^+, Cl^-–Cl^-, and K^+–Cl^-)
$N = n$ an ionized chlorine ion with a normal lattice ion (Cl^{+n}–Cl^- and Cl^{+n}–K^+); $(n + 1)$ electrons have been removed
$N = n^2$ two ionized chlorine ions (Cl^{+n}–Cl^{+n})

Calculations were carried out varying both the degree of ionization of the chlorine ions (between one and four) and the lifetime of these ions (each ionic state having a corresponding lifetime).

The general pattern emerging from a set of typical computations is illustrated. Figure 96 shows trajectories of the ions for the case of triple ionization up to a time of 8×10^{-13} sec, and the final relaxation of the lattice is not shown since it makes the diagram over-complicated. Figure 97 shows consecutive configurations of the lattice as the event progressed towards equilibrium. The results may be summarized as follows:

(1) An outward movement of all the positive ions in the central region takes place (this includes the K^+ as well as Cl^{+n} ions).

(2) Very pronounced $\langle 100 \rangle$ focusions are generated which transmit energy outwards many lattice spacings (in the triply ionized case the disturbance travels some 10 lattice parameters in 3×10^{-13} sec). This is a highly efficient means of dispersing a considerable proportion of the initial electrostatic energy.

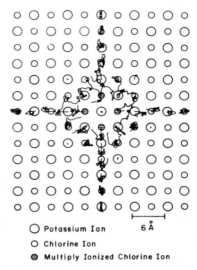

○ Potassium Ion   6 Å

○ Chlorine Ion

⊖ Multiply Ionized Chlorine Ion

Figure 96. Trajectories of the ions for the case of triple ionization of four chlorine ions up to a time of 8×10^{-13} sec.

The subsequent motion of the ions indicates an oscillation about mean positions displaced *outwards* from their original equilibrium lattice sites.

(3) The next nearest Cl⁻ neighbours move *inwards* under the attraction of the positive core. In the triply and quadruply ionized cases, some of these Cl⁻ ions move into the vacancy core left by the outgoing positive ions. This is illustrated for the triply ionized case (Figure 97(b)).

(4) There is a great deal of rearrangement of the central ions. For the case of triple ionization the following damage develops in $\sim 10^{-12}$ sec:

    (i)   Interstitials are produced by the outward expanding positive ions.

    (ii)   The Cl⁻ ions gather in the partially vacated central region left by the expanding core.

    (iii)   Ions interchange lattice sites.

(5) A secondary 'rearrangement explosion' occurs after electron capture as the central core, now predominantly filled with Cl⁻ ions, tries to regain its electrical neutrality (Figure 97(c) and (d)). A type of electrostatic annealing occurs. The ions seek to arrange themselves so that Cl⁻ ions settle on sites originally occupied by Cl⁻ ions, and similarly for K⁺. Thus, although some atoms have travelled as much as two or three lattice spacings, they come to rest on sites which are environmentally identical to those which they left. No visible damage remains.

These results of course take no account of the thermal vibrations initially present in the lattice, nor of the lattice 'thermal spike' due to electron/phonon interactions, which is estimated to be considerable in insulating compounds. It is therefore possible that a 'thermal ion explosion spike' could result in an outward thermal activation, and a more permanent production of stable positive ion interstitials, particularly with larger positive cores (produced by more energetic heavy particles).

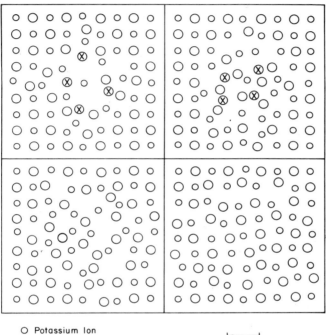

O Potassium Ion
◯ Chlorine Ion
⊗ Multiply Ionized Chlorine Ion

⊢————⊣
  6Å

Figure 97. Consecutive configurations of the central lattice as a function of time (triple ionization). (a) After 5×10^{-13} sec. Showing the outward movement of positive ions. (b) After 10^{-12} sec. This is just before the ions start their process of electron capture. Chlorine ions have moved into the partly vacant central core. (c) After $1 \cdot 5 \times 10^{-12}$ sec. The electrostatic annealing is now taking place. (d) After 3×10^{-12} sec. Ions have now rearranged themselves so that no visible damage remains. (Perturbations of ions from their lattice sites is due to their oscillations, which in turn transmits energy outwards.)

We must therefore conclude, at least from the present model, that the ion explosion spike alone is not sufficient to produce large numbers of point defects in ionic solids, since although damage occurs (4(i) and (ii) above) it is unstable and electrostatic annealing returns the lattice to normal after electron capture. In covalent solids, however, provided the lifetime criterion is satisfied, the thermal spike might thermally activate some of the ions participating in the ion explosion to more stable positions. A minimum time required for this would be $\sim 5 \times 10^{-13}$ to 10^{-12} sec, depending on the degree of ionization.

The Damage to Semiconducting Layer-structures

There *is* experimental evidence, however, for the existence of 'spikes' of electron excitation on a fission particle trajectory. Thus the fission fragment damage sustained by single crystals of molybdenum disulphide has been investigated using electron microscopy by several authors (Izui and Fujita, 1961, 1963; Chadderton, 1962a, b, 1964; and Caspar, 1963). Tracks produced by fission fragments in this material are normally observed as continuous black straight lines (the shorter tracks), or are intermittent (the longer tracks). The intermittent nature of the longer tracks has been proved to be real by observing that the discontinuities are preserved on tilting the crystals in a stereo-device (Chapter 6). In this section we consider the nature of the damaging mechanism and describe some experiments which substantiate a 'spike' model proposed for the production of intermittent tracks.

The Model

The majority of the energy lost by a fission fragment passing through a crystal is given to the electrons and, as we have seen, according to the thermal spike model these energized electrons (delta particles) diffuse outwards away from the trajectory of the fission fragment losing their energy in electron/electron and electron/lattice collisions. Consequently, the temperature of the lattice due to the electron excitation is expected to depend on the collision frequency for both these processes. A critical examination of these processes for the different crystal types accordingly leads to the conclusion that the lattice temperature due to the electron excitations is greatest in insulators, intermediate in semiconductors, and least in metallic crystals. Fission tracks due to electronic excitation may therefore only be formed if the electron mobility is less than some critical value which is characteristic of the material considered, hence establishing a resistivity dependence of track registration.

Molybdenum disulphide is one of a very large family of semiconductors which crystallize in a layer structure. They each contain basal planes of strong bonding and high atomic density, separated by relatively large interplanar channels where the bonding is mainly due to the weak Van der Waal forces. When a fission fragment passes through such a crystal at some angle α to the basal plane, it will cross successive basal sandwiches of high atomic density and through channel regions of low atomic density (Figure 98). Goland (1965) suggests that the rate of energy loss $(-dE/dx)$ will be greater along A–B than it will be along B–C; this is essentially a crystallographic energy-loss effect similar to channelling between the layers. Thus the energized electron density should be greater in the former region than in the latter. Now the total energy given to the lattice per unit length of the fission fragment path will also depend on the density of the delta particles excited. Therefore the lattice temperature due to the thermal spike will be greater when the fission particle travels in the dense 'sandwich' region than when it passes through the interplanar channel.

Consequently, intermittent tracks may be formed, provided that:

$$(-dE/dx)_{A-B} > (-dE/dx)_{critical} \qquad (13.1)$$

and
$$(-dE/dx)_{B-C} < (-dE/dx)_{critical} \qquad (13.2)$$

where $(-dE/dx)_{critical}$ is the critical rate of energy loss required for track formation. The angle α must also be small enough for the defect regions to be well separated and must be greater than a certain critical value so that

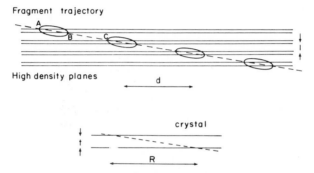

Figure 98. Model for the formation of intermittent tracks in layer structures.

more than one defect region is formed. As α increases the defect regions will merge to form a continuous track.

According to this model, therefore, the following predictions can be made regarding track registration in this class of semiconductors:

(1) Track registration should show some dependence on the electron conductivity of the crystals, and if the conductivity is sufficiently large no permanent tracks should be formed.
(2) Intermittent tracks will be characteristic of layer structures.
(3) The total number of intermittent segments comprising each track should be directly proportional to the specimen thickness.

Experimental

To investigate these predictions, fission fragment irradiation was carried out on a selection of six crystals with a similar layer structure but with varying resistivity (Table IX). The irradiation of these crystals presented the possibility of establishing a resistivity dependence of track registration when all other

15

factors were environmentally similar. The range of resistivities from molybdenum disulphide at one extreme to niobium ditelluride at the other is $\sim 10^6$ mΩ-cm.

Thin films of these compounds suitable for transmission electron microscopy were prepared by successively cleaving crystals along their basal planes using cellulose adhesive tape until they were reduced to a suitable thickness. They were then mounted on electron microscope grids and irradiated with fission fragments by placing the grids in contact with natural uranium foil $\sim 10\ \mu$ thick and introducing them for a specified time into the nuclear reactor. Post-irradiation observations were carried out in an AEI EM6 electron microscope (Morgan and Chadderton, 1968).

TABLE IX

Compound	Structure	Distance between layers (Å)	Electrical resistivity (25°C) (mΩ–cm)	Damage
MoS$_2$	MoS$_2$	6·13	$\sim 10^5$	Clearly defined tracks
MoSe$_2$	MoS$_2$	6·45	3×10^3	Tracks
WSe$_2$	MoS$_2$ (modified)	6·47	5×10^2	Tracks
WTe$_2$	MoS$_2$	6·27	2·8	Tracks
TaSe$_2$	MoS$_2$	6·37	4×10^{-1}	Ill-defined surface tracks
NbTe$_2$	related to CdCl$_2$	6·26	$2·8 \times 10^{-1}$	No tracks

Results

Tracks were formed in the materials with the larger resistivities (Table IX). However, between tungsten ditelluride (resistivity 2·87 mΩ-cm) and tantalum diselenide (resistivity 0·4 mΩ-cm) particle registration ceased. Thus for these semiconducting compounds tracks were evidently only formed when the resistivity $\gtrsim 1$ mΩ-cm (confirming hypothesis (1)), and in all those materials which were good registers of fission fragments the latter were of the same character as those formed in molybdenum disulphide (confirming hypothesis (2)).

Discussion

Consider an intermittent track making an angle α to the basal plane. Then, according to the model, provided that the crystal is perfectly flat the intermittency distance d will be a constant, and the product of the total number of spots N with the interlayer distance l will equal the foil thickness, $t(N \gg 1)$:

$$N.l \simeq t \qquad\qquad (13.3)$$

Thus, if a permanent visible defect is produced every time the fragment passes through a dense layer, then the measurement of the crystal thickness using (13.3) should agree with the thickness measured by some other independent method. Most crystals, however, are not planar. They suffer considerable and random bending when they are deposited onto the microscope grids. Hence for a particular track the intermittency may vary along its length, because α is now a function of position in the crystal. This may account for the gradual changes in track intermittency observed in bent crystals (Figure 29). An extreme example of this occurs when a fragment undergoes a close encounter with a lattice atom, where α changes abruptly, and when d will do the same. This is again found to be the case (Figure 32).

Consider a number of fission particle tracks in a relatively flat region of crystal. From similar triangles (Figure 98):

$$R = (t/l).d \qquad\qquad (13.4)$$

where: R is the length of track; d, average spacing between spots (d is only constant if α is a constant); t, the thickness of the crystal; and l is the distance between the sandwiches. If α is small (i.e. R is large) then d is large and defective regions are well separated and easily resolved. However, as R decreases, d decreases, and the individual regions get closer together. Their associated strain fields begin to interact, giving rise to a continuous line of contrast. After careful analysis of individual small tracks this fine structure has been observed even for tracks which initially seemed continuous. However, provided R is sufficiently large for individual dots to be resolved, equation 13.4 is obeyed.

In order to prove equation 13.3 a value for N is required. A thin crystal of molybdenum disulphide which contained a large number of intermittent tracks suitable for analysis was therefore selected. If such a crystal is of constant thickness (as might reasonably be expected for this type of cleaved crystal) then the independent values of N calculated from each track should be consistent with each other, and from these it is possible to calculate an average value of t. An average value of 420 ± 30 Å was obtained for t in this way, and this compared very favourably with the value of 450 Å calculated using the bent foil technique and observations of the electron diffraction pattern (Siems, Delavignette, and Amelinckx, 1962). The value is ~ 30 Å too small, but to some extent this smaller value was expected, since as the track approaches or leaves the specimen the damage mechanism is predominantly affected by the close proximity of the surface. The intermittent character is replaced by a short continuous *white* line and some uncertainty is introduced into the value of N. A similar consistency was achieved in four further specimens analysed – three molybdenum disulphide crystals and one tungsten diselenide (confirming prediction (3)).

Conclusion

On the basis of these observations the following conclusions may be drawn:

(1) The 'ion explosion spike' alone does not appear to be a sufficient model for the production of large numbers of point defects in fission fragment irradiated solids.

(2) The fission fragment tracks observed in certain semiconducting layer compounds are considered to arise from the agitation following the 'spike' of electron excitation produced along the fission fragment path. This model is supported by the resistivity dependence of track formation – a dependence predicted by theoretical considerations.

(3) A model is established which explains the appearance of intermittent tracks obtained in these structures. The intermittency is ascribed to the reduced energy loss when the fission fragment travels in the interplanar region.

(4) Analysis of the fine structure of the tracks confirms the presence of one defect aggregate per sandwich plane. The changes in the separation d of defect aggregates along the length of a track arise from a localized bending of the irradiated specimens, or from changes of particle direction due to collision events.

References

CASPAR, P. E. (1963), *Phil. Mag.* **8**, 96, 2091.

CHADDERTON, L. T. (1962a), *Nature* **195**, 987.

CHADDERTON, L. T. (1962b), *Proc. Roy. Soc.* **269**, 143.

CHADDERTON, L. T. (1964), *Proc. Roy. Soc.* **280**, 110.

CHADDERTON, L. T., MORGAN, D. V., and TORRENS, I. MCC. (1966), *Solid State Comm.* **4**, 391.

CHADDERTON, L. T., MORGAN, D. V., TORRENS, I. MCC., and VAN VLIET, D. (1966), *Phil. Mag.* **13**, 185.

FLEISCHER, R. L., PRICE, P. B., and WALKER, R. M. (1965), *J. Appl. Phys.* **36**, 3645.

GOLAND, A. N. (1965), Report BNL 9169.

IZUI, K., and FUJITA, F. E. (1961), *J. Phys. Soc. Japan* **16**, 1032; (1963), *J. Phys. Soc. Japan* **18**, 467.

MORGAN, D. V. (1966), Ph.D. Thesis, University of Cambridge.

MORGAN, D. V., and CHADDERTON, L. T. (1968), *Phil. Mag.* **17**, 1135.

NOGGLE, T. S., and STEIGLER, J. O. (1961), *Symposium on Radiation Effects*, American Society for Testing Materials.

SIEMS, R., DELAVIGNETTE, P., and AMELINCKX, S. (1962), *Phys. Stat. Solidi* **2**, 1, 421.

VARLEY, J. H. O. (1954), *Nature* **174**, 886.

CHAPTER FOURTEEN

Creation of Defects by Atomic Displacement

While most of the initial energy of a fission fragment is in fact passed to the electrons of a crystal we now know that it does not by any means follow that electronic excitation dominates in the creation of lattice defects. Indeed, in metals, where the excitation spreads over a large volume and quickly degrades, we are led to expect that the role played by direct atomic collision processes in defect production will be of major importance and that many of the phenomena revealed by computer simulation of a disturbed atomic array (Chapters 10 and 11) will play a crucial part. In this concluding chapter we therefore examine critically the experimental observations of the damage produced in gold (Chapter 7) and establish a model for the production of point defect clusters and extended line defects following atomic displacements in this particular metal (Van Vliet and Chadderton, 1967a, b).

Analysis of the Fission Fragment Spectrum

We first try to correlate the energy required to create clusters with the energy actually lost in the gold film by the bombarding fission fragments. Before making quantitative assessments it is essential to determine the exact mass-energy-angle spectrum for the fission fragments striking the specimens. This in turn requires an exact knowledge of the mass-energy spectrum of the particles created by the binary fission of a U^{235} nucleus. The mass spectrum has been well established (Chapter 1) and has two broad maxima centred about mass numbers of approximately 96 and 137. There is, however, no direct correlation between the mass and energy of fragments, and a considerable statistical variation has been discovered (Gibson, Thomas, and Miller, 1961). For the accuracy we require it will now be sufficient, as it was in Chapter 1, to consider that each fission event creates two fragments, a median-light and a median-heavy fission fragment. Following Brinkman (1960), we choose the atomic mass, nuclear charge, and initial energy as follows:

Median-light fission fragment	Median-heavy fission fragment
$M_L = 96$ amu	$M_H = 137$ amu
$Z_L = 39$	$Z_H = 55$
$E_L = 95$ MeV	$E_H = 67$ MeV

Since many of the fragments are created relatively deep within the uranium foil their energy is considerably attenuated when they actually emerge from

217

the surface. It may be shown that if the range of the fission products in uranium is less than the foil thickness the fragments emerge isotropically with an energy distribution given by:

$$N(E)_{\text{L,H}} = \rho\sigma\omega \, \frac{1}{4S(E)_{\text{L,H}}^{\text{U}}} = n(E)_{\text{L, H}} \, \Omega \tag{14.1}$$

where $N(E)_{\text{L, H}} \, dE$ is the total number of median-light or median-heavy fission fragments in the energy range $(E, E + dE)$ emerging per unit area per unit time, ρ is the number of U^{235} atoms per unit volume in the foil, σ is the cross-section for fission (taken as 500×10^{-24} cm^2), ω is the thermal neutron flux ($1\cdot2 \times 10^{12}$/cm^2 sec in BEPO) and $S(E)_{\text{L, H}}^{\text{U}} = (\partial E/\partial x)$ is a fraction for the total energy lost per unit length in uranium by a median-light or median-heavy fission fragment of energy E (the stopping power). $n(E)_{\text{L, H}} = 1/[4S(E)_{\text{L, H}}]$ is therefore the relative spectrum and $\Omega = \rho\sigma\omega$ is the number of fission events per unit volume. Now we already know that, neglecting radiation, $S(E)_{\text{L, H}}^{\text{U}}$ may be written as the sum of a nuclear and electronic stopping power (Chapter 2):

$$S(E)_{\text{L,H}}^{\text{U}} = S_{\text{n}}(E)_{\text{L,H}}^{\text{U}} + S_{\text{e}}(E)_{\text{L,H}}^{\text{U}} \tag{14.2}$$

The total fission fragment flux emerging from unit surface area per unit time, independent of mass, energy, and angle, is given by:

$$\phi = \frac{(R_{\text{L}}^{\text{U}} + R_{\text{H}}^{\text{U}})}{4} \, \Omega \tag{14.3}$$

where $R_{\text{L, H}}^{\text{U}}$ is the range of a median-light or median-heavy fission fragment in uranium.

The problem now is to calculate $S_{\text{nL,H}}^{\text{U}}$ and $S_{\text{eL,H}}^{\text{U}}$. For nuclear collisions we use a 'universal energy-loss plot' due to Lindhard, Scharff, and Schiøtt (1963). S_{nL}^{U} is plotted in Figure 99 for comparison with the results of the formula given by Bohr (1948) which is valid at high energies. Above 10 MeV the agreement is good. At lower energies (< 1 MeV), ranges calculated by the Lindhard plot agree well with experimental data (Davies, Brown, and McCargo, 1963). We shall accept his plot as being valid over the whole energy range.

No one theoretical approach to electronic stopping has yet given results which show a completely satisfactory agreement with experimental results (e.g. range measurements at high energies). However, most models do predict the same functional dependence of the electronic stopping power on energy (see Chapter 2), so that by knowing the ranges of fission fragments in uranium we can fit parameters to match the data.

Lindhard, in unpublished work, finds that, for velocities less than $v_1 = e^2 Z^{2/3}/\hbar$, S_{e} is proportional to the particle velocity, i.e. to $E^{1/2}$. For a median-light fission fragment $v_1 = 2\cdot6 \times 10^9$ cm/sec as compared to its initial velocity $v_0 = 1\cdot4 \times 10^9$ cm/sec. For a median-heavy fragment $v_1 = 3\cdot2 \times 10^9$ cm/sec and $v_0 = 0\cdot95 \times 10^9$ cm/sec. Hence the Lindhard theory should be applicable to both particles over their entire energy range.

An older formula due to Bohr (1948) yields for $v \ll 2Ze^2/\hbar$:

$$S_e \propto \frac{(Z_{eff})^2}{E} E^{1/2} \qquad (14.4)$$

where Z_{eff} is the net positive charge on a fast ion or fission fragment. This, as we have mentioned, is energy dependent since a fast-moving ion tends to drop some of the electrons in its outer shells. On the basis of a rather rough model, Bohr predicts $Z_{eff} \propto E^{1/2}$. Measurements by Fullmer and Cohen (1958) confirm this for fission fragments travelling in a gas. If this is also valid in a solid the Bohr formula gives $S_e \propto E^{1/2}$.

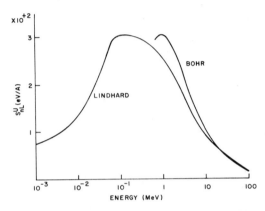

Figure 99. The rate of energy loss through direct nuclear collisions for a median-light fission fragment in uranium as given by Lindhard and colleagues, and by Bohr.

Furthermore, a purely empirical formula due to Kahn, Harman, and Forque (1965) which correlates the range-energy measurements of many different workers also predicts an $E^{1/2}$ dependence for S_e.

We assume that:

$$S_{e \, L,H}^{U} = K_{L,H}^{U} E^{1/2} \qquad (14.5)$$

where $K_{L,H}^{U}$ will be determined from experimental data. Niday (1961) has measured the mean ranges of fission products of different masses in uranium, and found a range of 6·07 μ for products of mass number 96, and 4·91 μ for those of mass number 137. Assuming that the initial energy values chosen here for median-light and median-heavy fragments correspond to the measured mean range, and using Lindhard's universal range-energy plot incorporating both electronic and nuclear stopping, we arrive at values of $K_{L,H}^{U}$ which yield:

$$S_{eL}^{U} = 2·85 \times 10^2 \, E^{1/2} \, \text{eV}/\text{Å} \qquad (14.6)$$

$$S_{eH}^{U} = 2·61 \times 10^2 \, E^{1/2} \, \text{eV}/\text{Å} \qquad (14.7)$$

where E is expressed in MeV.

The relative energy spectrum $n(E)_{L,H}$ has been calculated and is plotted in Figure 100. As one would expect from the energy dependence of S_e, $n(E)_{L,H}$ is strongly peaked at the lower energies. The mean energies of emergent median-light and median-heavy fission fragments are 30 and 27 MeV respectively, considerably less than the initial energies of 95 and 67 MeV.

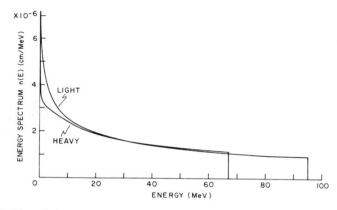

Figure 100. The relative energy spectrum of both median-light and median-heavy fission fragments emerging from a uranium foil of thickness greater than the range of either particle. The angular distribution is isotropic.

The Energy Loss Spectrum in Gold

In calculating the total energy lost by fission fragments in irradiated specimens it is necessary to consider the geometry of the specimen and the foil, since fragments travelling obliquely through the film 'see' a thicker film than those travelling normally. For a specimen lying flat on the surface of the foil the effective film thickness for all fragments is twice the actual thickness.

We can now make quantitative estimates of just how much energy is dissipated in the gold film by direct nuclear collisions and by electronic excitation, and just which energy ranges of the fission fragment spectrum are most instrumental in these two processes. We consider that the energy loss rates of the fission fragments are constant as they pass through the film. This approximation is not strictly valid for fragments whose ranges in gold are of the same order of magnitude as the film thickness, but this affects only a relatively small part of the spectrum. We now introduce two dimensionless functions of energy which we shall call the nuclear and electronic spectral damage functions and which are defined respectively as follows:

$$A(E)_{L,H} = n(E)_{L,H} \, S_n(E)^{Au}_{L,H} = \frac{S^{Au}_{n\,L,H}}{4(S^U_{e\,L,H} + S^U_{n\,L,H})} \qquad (14.8)$$

$$B(E)_{L,H} = n(E)_{L,H} \, S_e(E)^{Au}_{L,H} = \frac{S^{Au}_{e\,L,H}}{4(S^U_{e\,L,H} + S^U_{n\,L,H})} \qquad (14.9)$$

where $S_n(E)^{Au}_{L,H}$ and $S_e(E)^{Au}_{L,H}$ are the nuclear and electronic energy loss rates of median-light and median-heavy fragments in gold. From (14.1) and (14.8), $2A(E)_{L,H}\Omega dE$ is the energy lost per unit volume per unit time in the gold film through nuclear collisions by light or heavy fragments in the energy range $(E, E + dE)$. The factor 2 arises from the geometrical considerations discussed above. $2B(E)_{L,H}\Omega dE$ has the same meaning for electronic losses.

We also define the relevant energy integrals:

$$A_{L,H} = \int_0^{E_{L,H}} A(E)_{L,H} \, dE \qquad (14.10)$$

$$B_{L,H} = \int_0^{E_{L,H}} B(E)_{L,H} \, dE \qquad (14.11)$$

$$A = A_L + A_H \qquad (14.12)$$

$$B = B_L + B_H \qquad (14.13)$$

$2A_{L,H}\Omega$ is then the energy lost per unit volume per unit time through nuclear collisions by all light or heavy fission fragments; $B_{L,H}$, A, and B have similar meanings.

For the sake of consistency the stopping powers in gold must be chosen in the same way as those in uranium. Hence for $S^{Au}_{nL,H}$ we again use the Lindhard plot, and hope that some of its errors may be cancelled out, since in the definition of $A(E)_{L,H}$, $S^{Au}_{nL,H}$ and $S^{Au}_{nL,H}$ appear essentially as a ratio. $S^{Au}_{eL,H}$ is calculated by comparison with $S^{Au}_{eL,H}$ assuming that in general $S_e \propto n_0 Z^{1/2}$ (Kahn, Harman and Forque (1965)), where n_0 is the atomic density and Z the atomic number of the stopping medium. The Bohr and Lindhard formulae give slightly different dependences on Z, but owing to the close atomic numbers of gold and uranium (79 and 92) the discrepancy is small. This then gives:

$$S^{Au}_{eL} = 3\cdot24 \times 10^2 \, E^{1/2} \, eV/\text{Å} \qquad (14.14)$$

$$S^{Au}_{eH} = 2\cdot96 \times 10^2 \, E^{1/2} \, eV/\text{Å} \qquad (14.15)$$

where again E is expressed in MeV. We note that the maximum energy loss rates in gold are then 3160 eV/Å for a median light fission fragment and 2420 eV/Å for a median heavy.

$A(E)_L$ and $B(E)_L$ are plotted in Figure 101; similar curves are obtained for $A(E)_H$ and $B(E)_H$. In Figure 102 normalized histograms of $A(E)_L$ and $A(E)_H$ are shown so as to emphasize the dominant effect of relatively low energy fission fragments on the nuclear collision processes. A_L, A_H, A, B_L, B_H, and B have been found by numerical integration. The results are given in Table X.

The following conclusions may be drawn from these results:

(1) Almost all energy dissipated in the gold films arises from electronic excitation, and the loss is spread almost equally over the spectrum of fission fragment energies.

(2) The energy dissipated in nuclear collisions is correspondingly small (4%
for light fragments, 11% for heavy fragments, and 7% overall), but almost all
of this energy is from fragments which have already been severely attenuated
by the uranium foil. Therefore any proposed damage mechanism based on

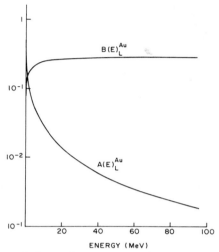

Figure 101. The nuclear and electronic spectral damage functions of median-light fission
fragments in gold, $A(E)_L$ and $B(E)_L$ respectively.

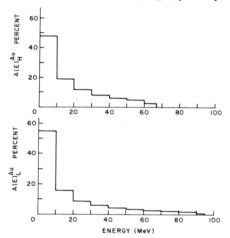

Figure 102. Normalized histograms of the nuclear spectral damage functions in gold of:
(a) median-heavy fission fragments; (b) median-light fission fragments.

nuclear collisions must be primarily concerned with those relatively low
energy fragments ($E < 10$ MeV).
(3) The energy below which nuclear collisions become the dominant energy loss
process is 800 keV for a median light fragment in gold and 1·3 MeV for a median

heavy fragment. These values are considerably higher than the generally accepted ionization thresholds of from 200 to 300 keV in gold (Dienes and Vineyard, 1957).

TABLE X

	Integrated spectral damage functions (MeV)	
	Median-Light	Median-Heavy
$A_{L,H}$	1·2	2·2
$B_{L,H}$	25·8	16·8
$A_{L,H} + B_{L,H}$	27·0	19·0

$A = 3\cdot4 \quad B = 42\cdot6 \quad A + B = 46\cdot0$

Changing the Energy Spectrum

The contribution of nuclear collisions to the visible damage might be assessed by screening out the high energy fission fragments, thus increasing the proportion of low energy fragments in the bombarding flux, and correspondingly increasing the proportion of energy lost through nuclear collisions. This may be done, for example, by coating the uranium foil with a layer of non-fissionable material to such a depth that almost all fission fragments are stopped within this layer. Those that do emerge have a severely attenuated energy. If the screening layer has exactly the same stopping power as the uranium foil, the effective energy spectrum $N'(E)$, and the effective spectral damage functions, $A'(E)_{L,H}$ and $B'(E)_{L,H}$, may be easily calculated:

$$N'(E)_{L,H} = \begin{cases} N(E)_{L,H}\left[1 - \dfrac{t^2}{L(E)_{L,H}^2}\right] & L_{L,H} > t \\ 0 & L_{L,H} < t \end{cases} \qquad (14.16)$$

$$A'(E)_{L,H} = \begin{cases} A(E)_{L,H}\left[1 - \dfrac{t}{L(E)_{L,H}}\right] & L_{L,H} > t \\ 0 & L_{L,H} < t \end{cases} \qquad (14.17)$$

$$B'(E)_{L,H} = \begin{cases} B(E)_{L,H}\left[1 - \dfrac{t}{L(E)_{L,H}}\right] & L_{L,H} > t \\ 0 & L_{L,H} < t \end{cases} \qquad (14.18)$$

where t is the thickness of the screening layer and $L(E)_{L,H}$ is the distance in which a median-light or median-heavy fission fragment slows down from its initial energy to energy E. These formulae incorporate an integration over the angular distribution (no longer isotropic since an energy dependent azimuthal cut-off angle exists), such that in calculating the total energy loss the effective film thickness is still twice the real value. The energy integrals, A'_L, etc., can then be compared directly to the former, A_L, etc., and retain exactly the same meaning.

$A'(E)_{L,H}$ and $B'(E)_{L,H}$ have been calculated for $t = 3$, 4, and 5 μ, and the corresponding values of A'_L, A'_H, B'_L, and B'_H found by numerical integration. The absolute results are given in Table XI and as percentages of the corresponding unscreened values in Table XII.

This experiment has in fact been performed using equivalent thicknesses of

TABLE XI

The effect of screening layers (μ) on the total energy losses (MeV)

Thickness	E_L^{max}	E_H^{max}	A_L	A_H	A	B_L	B_H	B	$A + B$
0	95	67	1·2	2·2	3·4	25·8	16·8	42·6	46·0
3	30	19	0·45	0·41	0·86	1·9	0·7	2·6	3·5
4	17	8	0·25	0·12	0·37	0·6	0·1	0·7	1·1
5	7	0	0·07	0	0·07	0·1	0	0·1	0·17

TABLE XII

Percentage decreases as a function of screening layers

	Thickness		
Quantity	3 μ	4 μ	5 μ
A'/A	26%	11%	2·1%
B'/B	6	1·6	0·2
$(A' + B')/(A + B)$	8	2·4	0·4
Effective dose	23	5·5	3

gold for the screening layer. The use of a screening medium of a slightly different stopping power changes the analysis in only a minor way. On the assumption that the stopping power of gold is 1·15 times that of uranium, gold layers 2·60, 3·47, and 4·34 μ thick (accurate to $\pm15\%$) were vacuum deposited onto natural uranium foil. Two different sets of 350 Å thick as-grown gold films were irradiated for 10 h in the BEPO reactor with natural uranium foil and with the three coated foils. One of these was specimen set A (see Chapter 7) and the other set was identically prepared, so that the dose-density dependence for no screening was well known. After irradiation the cluster density was measured in all cases, and the corresponding effective dose with no screening found from the observations described in Chapter 7. The average values of these doses expressed as a percentage of the no-screening dose are given in Table XII.

It would seem that a screening layer of 3 μ (or rather its equivalent) has roughly the same effect on cluster production as does lowering the total dose (without screening) by 23%. The total energy lost, in nuclear collisions is almost the same in both cases, whereas the total energy lost through electronic excitation is almost four times as great in the latter case. This difference is much more pronounced with the 5 μ thick screening layer. It would appear that the observed damage is far more, if not totally dependent on nuclear collisions than on electronic excitation. Although both the experimental results and the theoretical calculations are open to very large errors, the observed effect is sufficiently large that the interpretation must be qualitatively correct. We do not wish to imply that the concentration of visible clusters is directly proportional to the total energy lost through nuclear collisions, but rather that the cluster creating process is in some way critically controlled by the nuclear collision processes.

Damage from Nuclear Collisions

The question of possible damage arising from close nuclear collisions between fission fragments and the host gold atoms will be determined by the fate of the high energy primary knock-ons which are displaced from their lattice sites. These primaries of course are themselves slowed down by the crystal in the same manner as the fission fragments, through electronic excitation and elastic collisions with lattice atoms. The effect of the former on permanent damage is again negligible. As the primary travels through the crystal it produces a number of secondary displaced atoms through elastic collisions, and these in turn may displace even more atoms. The primary is, therefore, brought to rest in a spreading sequence of elastic collisions – the collision cascade. We need not consider the more general aspects of collision cascades since these have been dealt with elsewhere (Chapter 2; Chadderton, 1965; Holmes, 1964). It will be sufficient for our purposes to note that the immediate effect of collision cascades is the creation of equal numbers of interstitials and vacancies, and that the subsequent motion of these point defects determines the final defect structure.

Two basic mechanisms have been proposed for the formation of visible clusters in irradiated metals: depleted zone formation and the clustering of mobile point defects.

Depleted zones – localized vacancy-rich regions left near the end of the path of a high-energy primary or secondary – are supposed subsequently to rearrange themselves into visible clusters which are, of necessity, vacancy in character (Chapter 2). They are created at their final size and their formation is heterogeneous in nature.

Clustering occurs when mobile defects (either point defects or small complexes such as the di-vacancy) are trapped at stable nucleation sites where, by accumulating sufficient defects, small complexes grow to visible dimensions. Such clusters can clearly be of either a vacancy or interstitial character. Their growth rate is determined by the rate of creation and the mobility of point defects, as well as by the presence of other competing sinks. Nucleation sites can be created homogeneously by the chance meeting of two or more point defects to form a stable complex, or heterogeneously at impurity atoms or sub-microscopic depleted zones. The two mechanisms are thus not mutually exclusive; depleted zones may either grow or shrink by trapping migrating point defects.

The question of the final configuration of the clusters does not arise here – we are only concerned with how aggregates of a visible size may be formed. The actual structure (e.g. void, dislocation loop, stacking fault tetrahedra) is determined by the minimum energy configuration.

We shall discuss these two mechanisms in greater detail.

Clustering

The essential condition for clustering is that nucleation sites are available and that these absorb migrating defects faster than they evaporate point defects by thermal activation. To evaluate the possibility of clustering an exact knowledge of the diffusion rates of point defects and of the binding energies in complex clusters is essential. Since both interstitials and vacancies are mobile in gold during irradiation in the BEPO reactor ($T \sim 100°C$) aggregates of either type are clearly possible. The presence of a nearby insatiable sink, the film surface, affects the process critically.

Diffusion Coefficients

Vacancy diffusion in gold is now a farily well-understood process as a result of accurate quenching and annealing studies. De Jong and Koehler (1963) have shown that, under conditions of thermodynamic equilibrium, vacancies spend a finite proportion of their lifetimes combined in di-vacancies, and that vacancy diffusion is actually a combined diffusion of both single and di-vacancies. Since the activation energy for migration is less for a di-vacancy than for a single (0·66 versus 0·82 eV), the higher the vacancy concentration, the more time a vacancy spends combined in a di-vacancy, and the greater will be the effective diffusion coefficient. Thus, estimating the ambient vacancy concentration in the wake of a fission fragment as being of the order of 10^{-3}, we find, using the formulae and values given by de Jong and Koehler, that the effective vacancy diffusion coefficient should be about five times that for single vacancies at 100°C:

$$D_v^{eff} \sim 7 \times 10^{-13} \text{ cm}^2/\text{sec} \tag{14.19}$$

Interstitial migration in gold is still somewhat of an unknown quantity. We adopt here the view of Seeger (1962) that long-range interstitial migration in gold is due to dumb-bell interstitials which become mobile in Stage III ($\sim 0°C$). This is in accord with the experimental results of Venables and Balluffi (1965a, b) and of Bauer and Sosin (1964). The former found a value for the interstitial diffusion coefficient at $27°C$ of $1·7 \times 10^{-14}$ cm^2/sec, and associated with this an activation energy of $0·75 \pm 0·06$ eV (cf. $0·71 \pm 0·02$ eV by Schule, Seeger, Schumacher, and King (1962), and $0·80 \pm 0·04$ by Bauer and Sosin). Extrapolating these values to $100°C$ we find:

$$D_i = 5 \times 10^{-12} \text{ cm}^2/\text{sec} \tag{14.20}$$

The Lifetimes of Point Defects

The concentration of point defects within the lattice is determined by their rate of creation, and by the rate at which they diffuse out to the surface. Since a point defect at a distance x from a surface diffuses out in a time of approximately $x^2/4D$, the mean lifetime of point defects created uniformly within a film of thickness l is about $l^2/48D$. For $l = 350$ Å at $100°C$ the mean lifetimes of interstitials and vacancies are $\tau_i \sim 5 \times 10^{-2}$ sec and $\tau_v \sim 0·4$ sec.

An upper limit to the concentration of point defects may be set as follows. If each fission fragment creates its vacancy-interstitial pairs within a volume of cross section A in the plane of the film, then the average time, τ_f, between the overlap of two such sections is approximately $1/2A\phi$, where ϕ is the fission fragment flux. If we take A as a circle of diameter 1000 Å and $\phi = 6·0 \times 10^9$ FF/cm^2 sec (using enriched uranium), $\tau_f \sim 1$ sec. Using natural uranium, $\tau_f \sim 2$ min. Since τ_f is longer than either τ_i or τ_v then, if our assumptions are correct, point defects created by one fission fragment can diffuse out to the surfaces before another fission fragment passes through the same region.† Hence the maximum possible point defect concentration should be that created in the wake of a single fission fragment.

We note that τ_f has been severely underestimated. As was pointed out above, most of the damage created in nuclear collisions arises from low energy fragments, and the effective damage-creating flux is thus very much less than ϕ. The same conclusion will thus also hold true for thicker films, lower specimen temperatures, or for higher flux irradiations.

Interstitial Clustering

The conditions for homogeneous interstitial clustering have been considered by Venables and Balluffi (1965a, b). Once formed any multiple interstitial complex is essentially immobile. It grows by trapping migrating single interstitials

---

† This might well explain why the damage observed experimentally is independent of the fission fragment flux.

and shrinks by thermally activated emission of single atoms. An n-fold complex has a positive net rate of growth if:

$$V_n c_1 \exp\left(B_n/kT\right) > 1 \qquad (14.21)$$

where V_n is a geometrical factor, generally about 3, c_1 is the atomic concentration of single interstitials, and B_n is the binding energy of a single interstitial to an n-fold complex. The growth of visible interstitial clusters is only possible when this inequality is satisfied for all n.

A necessary condition is that the smallest possible multiple, the di-interstitial, has a positive growth rate. We estimate V_2 as 57/20 (57 and 20 are approximate co-ordination numbers for the absorption and emission processes respectively). Estimates of B_2 for gold vary considerably. Schumacher and Seeger (1963) give a value of 0·44 eV based on resistivity increases in Stage II annealing of cold-worked gold. Johnson's (1965) theoretical calculations for copper give $B_2 = 0·61$ eV, and we might reasonably expect a comparable value for gold. Venables and Balluffi (1965a, b), on the other hand, estimate B_2 as 0·11 eV, and this predicts a positive growth rate for di-interstitials at 100°C if $c_1 > 1 \times 10^{-2}$. If, however, $B_2 = 0·44$ eV, the critical concentration drops to 4×10^{-7}. How do these values compare with estimates of the interstitial concentration left in the wake of a fission fragment?

The interstitials created by a collision cascade in gold are liable to be widely dispersed owing to the frequent initiation of focused replacement sequences which transport interstitial atoms relatively long distances from their initial lattice sites within the cascade (Chapter 11). The mean distance, λ, of such sequences in gold has been estimated to be 100 Å at 10°C by Venables and Balluffi, 1000 to 2000 Å at 1·7°K by Swanson and Piercy (1964) (both indirectly from experimental work), and $\sim$30 Å at 70°C by Nelson, Thompson, and Montgomery (1962) on theoretical grounds. Now range measurements in gold indicate that the maximum possible energy loss rate of a gold primary knock-on in a gold lattice through elastic collisions is around 300 eV/Å. If 150 eV is required on the average to produce one vacancy-interstitial pair, then an energetic primary in gold produces on the average less than two interstitials per Å of path length. Since these are probably spread over a cylindrical volume with a radius of at least λ, the maximum atomic concentration of interstitials produced by one primary is then 9×10^{-3} for $\lambda = 35$ Å, 1×10^{-3} for $\lambda = 100$ Å, and 3×10^{-6} for $\lambda = 2000$ Å. These figures are overestimates in that they neglect the distance which a displaced atom travels before its energy drops below the focusing limit.

The expected interstitial concentrations are somewhat less than the critical concentration if we take $B_2 = 0·11$ eV, but well above it if we take $B_2 = 0·44$ eV. Which value should we accept as being the most reliable? Since the value of 0·11 eV of Venables and Balluffi was derived from the direct observation of interstitial clustering in gold, it might appear to be the 'best' value to use in the present case. Moreover, Venables and Balluffi found that an atomic concentration of greater than 3×10^{-3} was necessary to produce visible interstitial clusters at 27°C, and this critical concentration must surely increase with temperature. Thus it appears that under our experimental conditions interstitial clustering is not likely.

It should perhaps be noted that further experimental work by Thomas and Balluffi (1966) on the bombardment of gold foils with low energy gold ions in ultra-high vacuum indicates that the results of Venables and Balluffi – on which our model of interstitial clustering has been so far based – were strongly influenced by specimen contamination. Thomas and Balluffi find *no* evidence for the formation of interstitial clusters in gold at temperatures above −30°C, and their results are consistent with interstitials becoming mobile at temperatures well below the Stage III temperature range. These results, however, should not affect our basic conclusions on the possibility of interstitial clustering in gold films bombarded with fission fragments. If such clustering is unlikely at −300°C, it would be even more improbable in gold films bombarded at 100°C, despite the relatively high 'instantaneous' interstitial concentrations left by fission fragment induced collision cascades.

Vacancy Clusters

The critical condition for homogeneous vacancy clustering given by de Jong and Koehler (1963) is that the mobile di-vacancy is more likely to meet a vacancy and form a thermally stable tri-vacancy than to dissociate into two singles. This gives the condition that clustering is possible if:

$$\tfrac{16}{7} c_1 \exp\left[(E_m^1 + B_2 - E_m^2)/kT\right] > 1 \qquad (14.22)$$

where c_1 is the atomic concentration of single vacancies, E_m^1 and E_m^2 are the energies of migration for single and di-vacancies respectively (0·66 eV and 0·82 eV), and B_2 is the binding energy of a di-vacancy (0·10 eV). The form of the inequality is somewhat different from that of (14.21) since in this case the di-vacancy is the more mobile defect, whereas formerly the single interstitial was the important entity.

At 100°C, (14.22) becomes:

$$c_1 > 1 \times 10^{-4} \qquad (14.23)$$

and this concentration is almost certainly exceeded within collision cascades. We shall show later that concentrations up to 10^{-1} may exist. However, concentrations of this order only arise from situations in which depleted zones are found, so that the two models would therefore appear to reduce to one under these circumstances. Some distinction may nonetheless still be drawn between the two. A vacancy cluster formed within one collision cascade will be considered as being due to the clustering mechanism if its formation requires thermally activated migration at the ambient specimen temperature, and due to the depleted zone mechanism if it forms instantaneously independent of the specimen temperature. Although vacancy clustering is quite possible within collision cascades, we shall show in the next section that it is of secondary importance compared with depleted zones.

Depleted Zones

The idea of a depleted zone was first proposed by Seeger (1962) as an extension of Brinkman's concept of the displacement spike. Near the end of its path, a high-energy primary displaces almost every atom it passes, and these secondaries in turn create additional displacements within a few interatomic spacings of their own point of creation. A core of high vacancy concentration is formed and the displaced atoms come to rest in a surrounding interstitial-rich shell (Figure 8, Chapter 2). The energy remaining after the initial displacement cascade manifests itself as lattice thermal energy which is then dissipated by lattice diffusion. As envisaged by Brinkman (1954, 1956, 1960) the interstitial shell collapses inward during the lifetime of this 'thermal spike', and almost totally annihilates the vacancy core. The net result is a few isolated Frenkel pairs and perhaps a few small multiple defects. Seeger has proposed, however, that long-range transport of matter, primarily by focused replacement sequences, may carry most of the interstitials far enough away from their point of creation within the vacancy core that they reach sites which are stable against recombination during the ensuing thermal spike. Moreover long-range energy transport by non-replacement focusing sequences (focusons) spreads the initial cascade energy over a wider volume, producing a less intense thermal spike. Under these circumstances recombination is not complete and a large vacancy-rich core or 'depleted zone' remains (Figure 9, Chapter 2).

A vacancy cluster visible in the electron microscope may form by the collapse or rearrangement of a depleted zone into an energetically more favourable configuration. Such a process must involve the movement of single or small multiple vacancies either during the thermal spike or by ordinary thermal migration at the ambient temperature. In the latter case we associate the formation with the clustering mechanism. Since point defects are mobile some vacancy-interstitial pair recombination must occur. The degree of recombination depends on the relative dispersion of the two defects immediately after the cascade.

The eventual size of the cluster is basically determined by four factors:

(1) The initial energy of the primary or energetic secondary.
(2) The distribution of vacancies and interstitials left by the cascade.
(3) The extent, intensity, and duration of the thermal spike.
(4) The mobility of point defects.

The size distribution of these clusters is not only determined by the distribution of initial primary energies, but also by some form of statistical probability since both the distribution of point defects and their recombination are essentially random. The maximum possible size of a depleted zone does, of course, set an upper limit on the size of clusters. We shall now attempt to apply these considerations to the case of gold.

The spatial extent and distribution of point defects within a high-energy collision cascade in gold is difficult to predict. The basic problem is that the interatomic potential is not well known in the pertinent energy range of from 10 to 100 keV (see Appendix 4), and any theoretical calculations of the distribution depend critically on the potential. An estimate of the maximum energy of a primary knock-on capable of creating a depleted zone may be obtained by applying Brinkman's original criterion (1954) for displacement spike production, namely, that an ion initiates a spike when the mean-free-path between displacement collisions drops below the mean interatomic distance. Using the momentum approximation and the Born-Mayer potential derived in Appendix 4 (this is valid since the impact parameter is of the order of $r_0/2$) the critical energy, E_1, for displacement spike formation is 43 keV (cf. Brinkman's value of 80 keV). The number of vacancy-interstitial pairs in such a spike is approximately $E_1/2E_d^{eff}$. E_d^{eff} is an effective displacement energy which depends not only on how much energy is required to displace an atom but also on how effective focusing sequences are in dissipating energy in small-angle non-displacement collisions. For gold we estimate $2E_d^{eff} = 150$ eV, so that the maximum number of vacancies within a displacement spike is about 300. The same rough maximum should also apply to depleted zones.

An estimate of the vacancy concentrations possible within a cascade may be made from experimental data. If we accept that the range of a 10 keV primary in gold is about 30 Å (van Lint, Schmitt, and Saffredini, 1961), then $10,000/150 \sim 67$ vacancies should be distributed over a sphere of some 15 Å radius – a concentration of about 10^{-1}. Although this may be something of an overestimate it is apparent that very high vacancy concentrations are produced in gold, and that small vacancy multiples must be very common indeed. Moreover the interstitials are distributed over a sphere whose radius is perhaps a distance λ greater than the vacancy-sphere radius (see above), so that their concentration is much less than that of vacancies. If we accept that any interstitial created within the vacancy sphere immediately undergoes recombination, then about 1 in 35 pairs will recombine if $\lambda = 35$ Å, and 1 in 300 if $\lambda = 100$ Å. These figures are, of course, very approximate indeed but they do show that recombination is far from total.

The extent of the diffusion expected during the thermal spike depends strongly on the mean energy per atom, $\bar{E}$, in the initial thermal spike. During the lifetime of the spike, τ_s, a vacancy makes about n jumps:

$$n \sim g\tau_s \nu \exp(-E_m/k\bar{T}_s) \qquad (14.24)$$

where g is the co-ordination number and $\bar{E} = 2kT_s/3$. If $\tau_s \sim 10^{-11}$ sec, $\nu = 10^{13}$ sec^{-1}, and $g = 12$, then a vacancy makes more than one jump if $E_m < E_m^{crit} \sim 4\cdot6\,\bar{E}$. For example, in a spike with 10 keV divided equally within a 100 Å radius sphere, $E_m^{crit} \sim 0\cdot18$ eV, but, if the initial radius is only 60 Å,

$E_m^{crit} \sim 0.9$ eV. These values do not depend very sensitively on τ_s, and 10^{-11} sec is typical of the lifetimes generally associated with thermal spikes (Lehmann and Leibfried, 1963; Thompson and Nelson, 1962). The ordinary activation energy for vacancy migration, 0·82 eV, is somewhat greater than reasonable values of E_m^{crit}, but this overlooks three additional factors:

(1) The energy of migration of a di-vacancy is less; only 0·66 eV,
(2) The possibility of 'hot atoms' of energy much greater than $\bar{E}$ within the spike,
(3) Elastic interactions between closely separated defects may lower the effective energy of migration.

On this basis it does appear that vacancy migration may indeed be activated during the lifetime of the thermal spike and that a vacancy configuration of lower energy may form. This conclusion is in agreement with the experimental results of Howe, McGuire, and Gilbert (1966), who have observed black spots in gold and copper foils bombarded with 100 keV oxygen ions at temperatures near 4°K and which they associate with depleted zones. Since vacancy diffusion is virtually non-existent at these temperatures the migration appears to be activated by the thermal spike.

Correlated close-pair recombination with activation energies ranging from 0·05 to 0·6 eV (Bauer and Sosin, 1961) must also be favoured. However the effective range is only about 15 Å, so that recombination is probably still far from total. Random interstitial migration on the other hand may not be so likely even though its energy of migration (~0·75 eV) is less than that for vacancies. This may be attributed to two facts: interstitials form in the outer 'cooler' regions of the spike; and the co-ordination number for interstitial migration is only 1.†

Theory, therefore, predicts that depleted zones in gold may collapse into vacancy clusters, and since there may be as many as 300 vacancies in one zone, these should be visible. Qualitative considerations, moreover, show that for irradiation with fission fragments the formation of depleted zones is far more frequent than is the clustering of single vacancies.

The differential cross-section, $d\sigma(E, T)$, for energy transfer, T, in an elastic collision between a fission fragment of energy E and a gold atom may be represented by:

$$d\sigma(E, T) \propto \frac{dT}{T^m} \qquad (14.25)$$

where m depends on the energy of the fragment and varies from 2 at very high energies (Rutherford scattering) to almost 0 at very low energies (hard sphere scattering). If the number of atoms displaced by a primary knock-on is

---

† It is assumed that long-range crowdion motion in gold does not occur.

directly proportional to its energy, then the total number of vacancies created by primaries of energy between T_1 and T_2 has a functional dependence:

$$n(T_1, T_2) \propto \begin{cases} \log_e (T_1/T_2) & m = 2 \qquad (14.26) \\ T_1^{-m+2} - T_2^{-m+2} & m < 2 \qquad (14.27) \end{cases}$$

There is a lower limit, T_{2L}, to the energy of primary knock-ons which can create stable clusters from depleted zones since very small clusters may be unstable or even mobile at the ambient temperatures. Thus, primaries of energy between the displacement energy, E_d, and T_{2L} create isolated vacancies, while those between T_{2L} and T_1 create depleted zones. Choosing extreme values of $E_d \sim 35$ eV, $T_{2L} \sim 1$ keV, and $T_1 \sim 43$ keV, and considering only primaries of energy less than T_1, we find that for $m = 2$ slightly more vacancies are created within depleted zones than are created as isolated vacancies, while for $m = 0$ only 1 in 2000 is isolated. The fission fragments which cause the most damage are those of energy below 10 MeV, and it is in this range that the screening of the nucleus by the inner electrons (see Appendix 4) becomes appreciable. The effect of the screening is to give m some intermediate value between 0 and 2, thus increasing the proportion of vacancies within depleted zones. A similar analysis applies to primaries of energy greater than T_1. These primaries must have a lower value for m than the fission fragments which displaced them, so that depleted zone formation by their secondary knock-ons is even more dominant. It therefore appears that depleted zones account for a much larger proportion of the total number of vacancies formed during irradiation than do isolated singles or small complexes. Since they are also much more effective in producing visible clusters, it follows that the vast majority, if not all, of vacancy clusters are actually depleted zones. Since little or no interstitial clustering is to be expected, we conclude that they comprise virtually all of the visible damage.

Correlation with Experiment

Collapsed depleted zones should have the following characteristics:

(1) There is a maximum cluster size associated with the maximum energy of a primary or secondary which can create a depleted zone.
(2) If the growth of pre-existing clusters is negligible, the distribution of sizes below the maximum is independent of dose.
(3) The concentration of clusters is directly proportional to the number of energetic primary and secondary knock-ons within a certain energy range, and is hence proportional to the total dose.
(4) The concentration must therefore also be independent of the film thickness, irradiation temperature, and flux.
(5) Denuded areas do not exist since no long-range migration of point defects is either invoked or required.

Conversely, clusters formed by the agglomeration of single point defects should present the following characteristics:

(1) The size increases steadily with dose.

(2) The concentration may or may not be directly proportional to the total dose, depending on whether the nucleation of new clusters and the growth of pre-existing clusters are affected by the total concentration.

(3) The concentration is strongly influenced by the film thickness, irradiation temperature, and flux.

(4) Denuded areas near sinks, especially the surfaces, must be prominent.

The clusters observed at low fission fragment doses exhibit the characteristic properties predicted for collapsed depleted zones (Chapter 7). Their concentration is directly proportional to dose, and independent of the film thickness and flux. The mean image diameter and distribution are only slightly dependent on the dose, and indicate that some growth of existing clusters does occur. Neither denuded areas near lattice defects nor large denuded surface layers are found. A slight denuded layer near a surface is of course to be expected since clusters which intersect a surface should anneal out immediately. These observations contradict most of the properties of the clustering process.

Depleted zones also explain qualitatively the groups of closely spaced clusters observed at relatively low doses (Figure 47, Chapter 7). These may be due to several energetic primaries or secondaries all arising from the same fission fragment and all of which create distinctly separated depleted zones.

A good quantitative correlation may also be drawn between experimental observations and the depleted zone theory. The maximum theoretical cluster size is about 300 vacancies. A Frank loop of 300 vacancies has a diameter of 50 Å, and the diameter of its image is of necessity somewhat larger. The largest observed images had diameters of around 100 Å, corresponding to 1100 point defects. It is possible that these clusters have not been formed by a single depleted zone, but by the coalescence of two or more adjacent zones. Thus the discrepancy in size does not appear to be too serious.

Moreover, a good correlation is found between the concentration of visible clusters and the expected number of primary knock-ons above a certain energy. The mean cross-section for elastic collisions, in which greater than 10 keV is given to a gold lattice atom, has been found by calculating the cross-sections for median-light and median-heavy fission fragments as a function of their energy and averaging over the energy distribution. In calculating the cross-sections, the momentum approximation and the screened Coulomb potential were employed. Applying the result to a 350 Å thick film, the number of primaries of energy greater than 10 keV is 0·8 per incident fission fragment. If the number of secondary and tertiary knock-ons of energy greater than 10 keV were also included, this figure would be somewhat increased. Experimentally, 0·6 visible clusters per incident fission fragment are observed. A depleted zone cluster formed by a 10 keV knock-on contains up to 65 vacancies or, as a

Frank loop, it has a diameter of less than 25 Å. The minimum visible image diameter was about 20 Å. It is thus certainly very enticing to associate each cluster of a visible size with a depleted zone formed by a primary or secondary of energy greater than about 10 keV. The agreement is strikingly good, but if we take into account the approximations and uncertainties involved in both the theoretical and experimental values it is possibly somewhat fortuitous.

The product of the energy spectrum and cross-sections for energy transfers greater than 10 keV is plotted as a function of energy in Figure 103 for median-light and median-heavy fission fragments in gold. Almost all such knock-ons

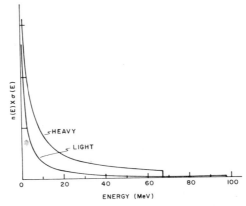

Figure 103. The product of the relative spectrum, $n(E)_{L,H}$, and the cross-section, $\sigma(E)_{L,H}$, for energy transfer greater than 10 keV in a direct nuclear ('elastic') collision between a gold lattice atom and an incident median-light or median-heavy fission fragment. The result is expressed in arbitrary units.

are due to the low-energy fragments. A similar effect has already been noted for the energy lost through all nuclear collisions and found to agree qualitatively with experimental observations. Thus the depleted zone mechanism does not contradict the former results.

At high doses we cannot expect that a theory which involves the creation of isolated clusters only will be sufficient. Complications must arise owing to the multiple overlapping of damaged regions, and the elastic interactions between the clusters themselves. A detailed investigation and consideration of such processes has not been made.

The explanation most consistent with both theory and experiment is that the clusters observed in fission fragment irradiated gold films are vacancy-type, collapsed depleted zones.

Conclusion

The examination using electron microscopy of thin, single-crystal, gold films bombarded with energetic uranium-235 fission fragments has revealed the presence of small defect clusters – the 'black spots' common to most irradiated metals. These have been interpreted as being exclusively vacancy clusters which are directly associated with depleted zones – vacancy-rich regions formed '*in situ*' by energetic primary or secondary knock-ons which arise from nuclear (elastic) collisions between the incident fragments and the lattice atoms. In choosing on a theoretical basis this particular model for damage we have rejected two other possible mechanisms: defect formation resulting from electronic excitation, and the agglomeration of point defects into visible clusters. In the former process, despite the considerable energy losses by the incident fission fragments in electronic excitation, this energy is too efficiently dissipated by the conduction electrons to leave any permanent trace. In the latter process, on the one hand the escape of interstitials to the surface sinks is too rapid to allow the build-up of an atomic concentration of interstitials sufficiently high to render the agglomeration process possible, while on the other hand the fraction of vacancies created outside depleted zones is judged to be small enough that the effects of vacancy clustering should be negligible compared with the production of depleted zones.

On the basis of this investigation into fission fragment damage in gold it is possible to make further predictions on fission fragment damage in thin films of the other metals. It has been noted that the damaging effects of electronic excitation should be negligible in all metallic crystals of high conductivity. Assessing the effects due to direct nuclear collisions in different metals is an extremely complex problem and no doubt depends very strongly on both specific properties of the metal concerned and on many parameters associated with the irradiation conditions. For example, although it appears that the effects of clustering have been unimportant in the investigation using gold, this certainly cannot be accepted as a general rule. We shall examine a few particular cases.

In copper and silver, point defects behave in a similar way to those in gold, and we expect that under the same irradiation conditions we would also observe clusters characteristic of depleted zones. In copper, however, the ions are much smaller, compared to the interatomic separation, than they are in gold, so that depleted zones and hence the visible clusters should be correspondingly smaller. Makin, Whapham, and Minter (1962) have associated images of diameter less than 25 Å observed in neutron irradiated copper with depleted zones, and such clusters should be typical of those in copper films bombarded with fission fragments. In irradiated silver the clusters should be of an intermediate size. In both cases the rate of cluster formation should be comparable to that in gold. No experimental work on fission fragment damage in copper or silver has yet been reported.

In aluminium, another metal of high point defect mobility, the ions are so small compared to the interatomic separation that the maximum size of a collapsed depleted zone is probably below resolvable dimensions. Thus any visible clusters must be formed by an agglomeration process. This implies that clusters of either vacancy or interstitial character may be present, and that their concentration and size must be dependent on both the temperature and fission fragment flux. Moreover, their rate of formation should be very much less than that observed in gold since most point defects are lost to the surfaces rather than trapped in clusters. These predictions are borne out by the experimental observations of Westmacott, Roberts, and Barnes (1962) on fission fragment damage in aluminium. The observed rate of formation of visible clusters is in fact several orders of magnitude less than that found in gold. Moreover they identify both vacancy and interstitial dislocation loops, and attribute their growth rate primarily to a thermally activated coalescence of loops during the irradiation.

In metals of relatively low point defect mobility it appears that there may be insufficient thermal activation, during the thermal spike, to trigger the collapse or re-arrangement of depleted zones. However, Ruedl, Delavignette, and Amelinckx (1962) have interpreted clusters in platinum irradiated with fission fragments as being essentially due to depleted zones, suggesting that collapse is still possible even when the energy of migration for a single vacancy is as high as 1·43 eV (Piercy, 1960). This in turn indicates that large depleted zones must be formed in almost all irradiated heavy dense metals at a rate comparable to that in gold. However, their formation is no doubt affected by the high background concentration of immobile point defects. Thus the formation rate found by Ruedl *et al.* in platinum appears to be comparable to that in gold, although no exact figures are given. They also find that, unlike gold, the size of the clusters increases with dose, suggesting perhaps that some of the otherwise immobile vacancies have migrated to depleted zones.

References

BAUER, W., and SOSIN, A. (1961), *Phys. Rev.* **136**, A255.

BAUER, W., and SOSIN, A. (1964), *Phys. Rev.* **136**, A474.

BOHR, N. (1948), *Math.-fys. Medd.* **18**, 8.

BRINKMAN, J. A. (1954), *J. Appl. Phys.* **25**, 961.

BRINKMAN, J. A. (1956), *Amer. J. Phys.* **24**, 246.

BRINKMAN, J. A. (1960), Rendiconti della Scuola Internazionale di Fisica 'E. Fermi', XVIII Corso, 830.

CHADDERTON, L. T. (1965), *Radiation Damage in Crystals*, Methuen (London).

DAVIES, J. A., BROWN, F., and MCCARGO, M. (1963), *Can. J. Phys.* **41**, 829.

DIENES, G. J., and VINEYARD, G. H. (1957), *Radiation Effects in Solids*, Interscience (New York).

FULLMER, C. B., and COHN, B. L. (1958), *Phys. Rev.* **109**, 94.

GIBSON, W. M., THOMAS, T. D., and MILLER, G. L. (1961), *Phys. Rev. Letters* **7**, 65.

HOLMES, D. K. (1964), *The Interaction of Radiation with Solids*, North-Holland, p. 147.

HOWE, L. M., MCGUIRE, J. F., and GILBERT, R. W. (1966), *Acta Met.* **14**, 801.

JOHNSON, R. A. (1965), *J. Phys. Chem. Sol.* **26**, 75.

JONG, M. DE, and KOEHLER, J. S. (1963), *Phys. Rev.* **128**, 40.

KAHN, S., HARMAN, R., and FORQUE, V. (1965), *Nucl. Sci. Eng.* **23**, 8.

LEHMANN, C., and LEIBFRIED, G. (1963), *Z. Physik.* **172**, 465.

LINDHARD, J., SCHARFF, M., and SCHIØTT, H. E. (1963), *Math.-fys. Medd.* **33**, 14.

LINDHARD, J. (1966), *Notes on Atomic Collisions IV*, unpublished.

LINT, V. A. J. van, SCHMITT, R. A., and SUFFREDINI, C. A. (1961), *Phys. Rev.* **121**, 1457.

MAKIN, M. J., WHAPHAM, A. D., and MINTER, F. J. (1962), *Phil. Mag.* **7**, 285.

NELSON, R. S., THOMPSON, M. W., and MONTGOMERY, H. (1962), *Phil. Mag.* **7**, 1385.

NIDAY, J. B. (1961), *Phys. Rev.* **121**, 1471.

PIERCY, G. R. (1960), *Phil. Mag.* **5**, 201.

RUEDL, E., DELAVIGNETTE, P., and AMELINCKX, S. (1962), *J. Nucl. Mat.* **6**, 46.

SCHULE, W., SEEGER, A., SCHUMACHER, D., and KING, H. (1962), *Phys. Stat. Sol.* **2**, 1199.

SCHUMACHER, D., and SEEGER, A. (1963), *Phys. Letters* **7**, 184.

SEEGER, A. (1962), *Radiation Damage in Solids*, IAEA (Vienna), **1**, 101.

SWANSON, M. L., and PIERCY, G. R. (1964), *Can. J. Phys.* **42**, 1605.

THOMAS, L. E., and BALLUFFI, R. W. (1966), *Appl. Phys. Letters* **9**, 171.

THOMPSON, M. W., and NELSON, R. S. (1962), *Phil. Mag.* **7**, 2015.

VAN VLIET, D., and CHADDERTON, L. T. (1967), *Phil. Mag.* **16**, 275.

VAN VLIET, D., and CHADDERTON, L. T. (1967), *Phil. Mag.* **16**, 291.

VENABLES, J. A., and BALLUFFI, R. W. (1965), *Phil. Mag.* **11**, 1021.

VENABLES, J. A., and BALLUFFI, R. W. (1965), *Phil. Mag.* **11**, 1039.

WESTMACOTT, K. H., ROBERTS, A. C., and BARNES, R. S. (1962), *Phil. Mag.* **7**, 2035.

Field-Ion Microscopy of Fission
Fragment Damage in Tungsten

The technique of field-ion microscopy is being used increasingly for the study of radiation damage. The microscope possesses two important advantages; firstly, a resolution of 2 to 3 Å enabling individual atoms to be distinguished, and, secondly, that it is possible by means of field evaporation to strip individual atoms or whole atomic layers from the surface under observation in a controlled manner and obtain a true three-dimensional picture of the damage. Accounts have been published of the investigation of the damage produced by neutrons (Bowkett, Hren, and Ralph, 1964; Müller, 1963), α-particles (Brandon, Southon, and Wald, 1962; Müller, 1960) and neutral atoms (Sinha and Müller, 1964; Brandon, Wald, Southon, and Ralph, 1963). The microscope may, however, be used to observe the damage produced by fission fragments (Bowkett, Chadderton, Norden, and Ralph, 1965, 1967).

The specimen used in the field-ion microscope takes the form of a fine wire electropolished to produce a tip (which is the region of the specimen that is imaged) having a radius of 100 to 500 Å. The simplest material to use as a specimen and the one which gives some of the best results is tungsten.

The experimental arrangement for the fission fragment irradiation of the specimen tips is shown in Figure 104. The size of the capsule is restricted by the irradiation facilities available in the reactor (BEPO), and the geometry was chosen to give the best possible collimation of the fission fragments. In fact the angular spread of particles for any one specimen was less than 7°. The field-ion specimens are mounted in this assembly in the same way that they are normally mounted in the microscope; that is spot welded onto a nichrome wire which is coiled around two supporting tungsten electrodes (in the case of the microscope), or two aluminium wire pins (for irradiation). Specimens may be annealed in the microscope by passing a current through the heater wire.

A small piece of enriched uranium foil is situated about 6 mm from the specimen tips and when the capsule is in the reactor the neutron bombardment induces fission and some of the fission fragments produced are emitted from the surface of the foil. Enriched uranium foil (90% ^{235}U) was used in order to minimize neutron damage in the specimens and the capsule was sealed under vacuum since the range of fission fragments in air is comparatively small. The thickness of the uranium foil used was 0·001 in., which is greater than the mean-free-path of the fission fragments and results in the energy spectrum of the fission fragments having a single peak about 30 MeV (Chapter 14, page 220).

The estimated total neutron dose on the specimens was 5×10^{14} nvt. The fission fragment dose was monitored by means of molybdenum trioxide crystals sublimed onto carbon-coated electron microscope grids which were mounted alongside the field-ion specimens as shown in Figure 104. A fission fragment passing through a molybdenum trioxide crystal registers as a hole which may be observed in the electron microscope and each individual hole

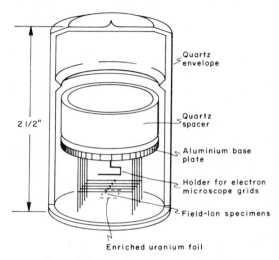

Figure 104. Experimental irradiation rig.

represents the passage of a single particle (Chapter 5, page 60). Ideally one and only one fission fragment should strike each specimen, and this condition may be achieved as closely as is possible within the limits of a statistical distribution. In a typical electron micrograph there was on average one hole in every 8×10^{-12} cm² whilst the area of cross-section of a typical field-ion tip is 6×10^{-12} cm².

Where a fission fragment strikes the surface of a tungsten specimen, material is evaporated and the damage observed is a crater which appears very strikingly in the field-ion image as shown in Figure 105. By field evaporation, atomic layers may be stripped from the surface and the depth of these cavities can be found directly. The pattern of damage within the specimen may also be investigated. Figure 105 is typical of how a specimen first appears once the four to five layers of oxide and impurity on the surface have been field evaporated. The very large central crater was about 250 Å radius, whilst the specimen surface itself was about 500 Å radius. The shape of the cavity was found to be approximately hemispherical by field evaporating the specimen layer by layer and comparing micrographs after evaporation of each layer.

105

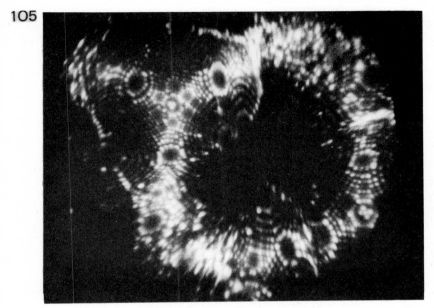

Figure 105. Field-ion micrograph of tungsten specimen showing large surface crater.

106

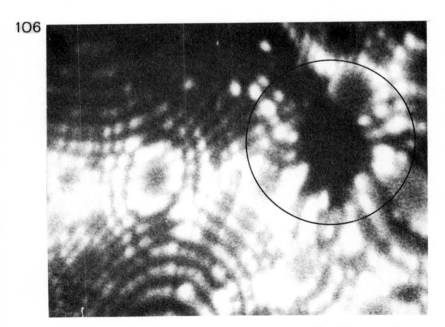

Figure 106. 35 Å diameter sub-surface cluster revealed by field evaporation.

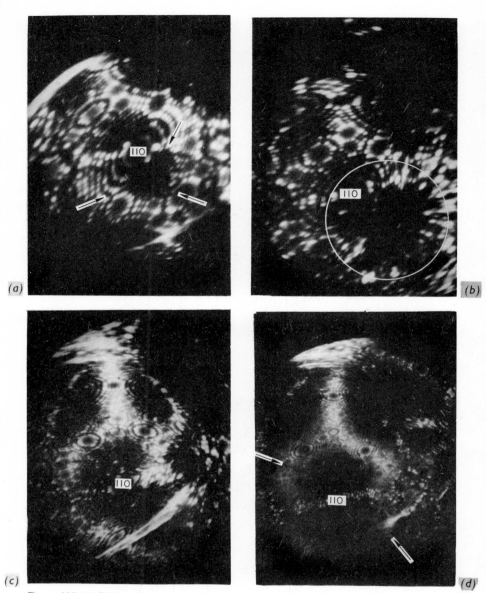

Figure 107. (a) Fission fragment irradiated tungsten specimen. Many layers have been field evaporated from the surface which still has a somewhat disturbed structure (note especially the dark region on the central (110) plane where the atoms are building down into a sub-surface vacancy cluster). (b) After the evaporation of two atomic planes from the surface depicted in (a) the vacancy cluster is revealed. The position of the crater is marked by the white circle, which has a diameter about twice that of the crater. (c) The same tungsten specimen as (b) after evaporating a further 45 atomic planes, the damage region is healing and moving across the specimen surface. (d) The same as (a) after evaporating a further 130 planes. The damage has nearly healed, but there is a grain boundary running across the (110) plane, with a misorientation of about 12°.

Fission fragments producing cascade events some distance below the surface might be expected to result in the occurrence of vacancy clusters having a size considerably smaller than the cavities at the surface. Such clusters are in fact uncovered by field evaporation. Figure 106 shows a section through a cluster of this type which had a diameter of 35 Å. Isolated single vacancies and a few small clusters which were detected could have been generated either by fission fragments or by neutrons, since the specimens also received a neutron dose of 5×10^{14} nvt. Separate investigations (Bowkett, Hren, and Ralph, 1964) on neutron irradiated tungsten wire which had received a comparable dose have shown that the clusters due to neutron damage will be nothing like as large as the one shown in Figure 106 which can thus be attributed unambiguously to fission fragment damage. Craters of these dimensions were never observed in

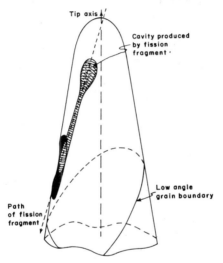

Tip axis

Cavity produced by fission fragment

Low angle grain boundary

Path of fission fragment

Figure 108. A schematic representation of the damage induced by the fission fragment as it passed through the specimen.

the tungsten used for these experiments when it was examined either un-irradiated or neutron irradiated. However, very occasionally, poorly prepared tungsten specimens can give rise to surface holes which bear a superficial resemblance to fission fragment induced damage.

When the irradiation geometry is altered so that the direction of collimation of the fission fragments makes an angle of about 20 °with the specimen axis, the damage revealed by field-ion microscopy is significantly different from that occurring in the axial case. Figure 107 shows a field evaporation sequence through a specimen irradiated in this manner. The region of dark contrast in Figure 107(a) is interpreted as a surface depression where the surface atoms are collapsing inwards towards a sub-surface vacancy cluster. Evaporation of a few atomic planes revealed this cavity (Figure 107(b)) as a true void rather than a dispersed region with a high vacancy concentration. Further evaporation of atomic layers showed that the cavity was pear-shaped with a long tail which finally emerged at the side of the specimen (Figures 107(c) and (d)). This event appears to be very similar to 'white' fission fragment tracks (Chapters 4 and 5) caused by removal of some of the crystal by the particle and revealed by electron microscopy. In the case of the non-axial bombardment of the tungsten tip, the fragment track emerges from the side of the specimen and material may be removed. The translation of the damage region normal to the axis during field evaporation allows us to determine the path of the fragment and magnitude of its effect on the surrounding lattice. Figure 108 illustrates schematically the observed damage in the form of a three-dimensional model, the incidence angle as deduced from the model being in agreement with the angle of collimation of the fragment within the experimental error.

It is interesting to note the appearance of a grain boundary in Figure 107(d), close to the point of exit of the fragment, although it is impossible to determine whether this was actually created by the asymmetry in field stress produced by the damage, or whether it was originally present in the tip.

The absence of the large surface craters seen following an axial bombardment may be explained by the fact that in the non-axial case the fragments have sufficient energy to pass through the specimen, removing material from the point of exit. An axial fission fragment must come to rest in the specimen and consequently the reflection of much of its momentum will tend to cause removal of surface atoms at the point of entrance of the fragment.

References

BOWKETT, K. M., CHADDERTON, L. T., NORDEN, H., and RALPH, B. (1965), *Phil. Mag.* **11**, 651.

BOWKETT, K. M., CHADDERTON, L. T., NORDEN, H., and RALPH, B. (1967), *Phil. Mag.* **15**, 415.

BOWKETT, K. M., HREN, J. J. A., and RALPH, B. (1964), *Third European Regional Conference on Electron Microscopy*, Prague, p. 191.

BRANDON, D. G., SOUTHON, M. J., and WALD, M. (1962), *Properties of Reactor Materials and the Effects of Radiation Damage*, Butterworths (London), p. 113.

BRANDON, D. G., WALD, M., SOUTHON, M. J., and RALPH, B. (1963), *J. Phys. Soc. Japan*, Suppl. II, **18**, 324.

MÜLLER, E. W. (1960), *Proceedings of the 4th International Conference on the Reactivity of Solids*, Elsevier (Amsterdam), p. 862.

MÜLLER, E. W. (1963), *J. Phys. Soc. Japan*, Suppl. II, **18**, 1.

SINHA, M. K., and MÜLLER, E. W. (1964), *J. Appl. Phys.* **35**, 126.

The Energy Interchange between Electrons and Lattice

This problem has been considered in some detail for the case of uranium by Ozeroff (1949) and Brooks (1950). The treatments which they give, however, can be applied more generally to insulators, metals, and semiconductors. The model is idealized and is treated classically.

A Maxwellian gas of electrons at a temperature T_1 is homogeneously mixed with a second Maxwellian gas of heavy atoms at a lower temperature T_2. This then represents the state of affairs which is assumed to prevail after passage of the fission fragment. The rate of energy interchange between the electrons themselves, and between the gas atoms, is assumed to be large compared with the rate of interchange between electrons and gas atoms, so that each of the component gases is maintained at equilibrium. The energy interchange may then be computed by considering a pair of particles, one of each type. In the centre of mass system it is:

$$I(\gamma) = \frac{3m_1 m_2}{(m_1 + m_2)^2} (kT_1 - kT_2)(1 - \cos \gamma) \qquad (A2.1)$$

where γ is the angle of scattering and m_1 and m_2 are the relevant masses. If in addition to averaging over the two Maxwellian velocity distributions we also average over all values of γ, assuming isotropic scattering, we obtain:

$$(\Delta E)_{av} = \frac{3m_1 m_2}{(m_1 + m_2)^2} (kT_1 - kT_2) . \qquad (A2.2)$$

The temperatures T_1 and T_2 then satisfy the equations:

$$\frac{d}{dt} (\tfrac{3}{2}kT_1) = - \frac{3m_1 m_2}{(m_1 + m_2)^2} \frac{1}{\tau} (kT_1 - kT_2) \qquad (A2.3)$$

$$\frac{d}{dt} (\tfrac{3}{2}kT_2) = \frac{3m_1 m_2}{(m_1 + m_2)^2} \frac{1}{\tau} (kT_1 - kT_2) \qquad (A2.4)$$

where τ is the time between electron/atom collisions. This may be written:

$$\tau \frac{(m_1 + m_2)^2}{4m_1 m_2} \frac{d}{dt} (\tau_1 - \tau_2) = -(\tau_1 - \tau_2) \qquad (A2.5)$$

so that the relaxation time is:

$$\tau_R = \frac{(m_1 + m_2)^2}{4m_1 m_2}\, \tau \simeq \frac{m_2}{4m_1}\, \tau \tag{A2.6}$$

The problem of coupling may be represented by the pair of differential equations:

$$K_1 \frac{\partial^2 T_1}{\partial x^2} = C_1 \frac{\partial T_1}{\partial t} + b(T_1 - T_2) \tag{A2.7}$$

$$K_2 \frac{\partial^2 T_2}{\partial x^2} = C_2 \frac{\partial T_2}{\partial t} - b(T_1 - T_2) \tag{A2.8}$$

where K_1 and K_2 are conductivities of electrons and atoms respectively, C_1 and C_2 are heat capacities per unit volume, and b is a coefficient of heat transfer between electrons and atoms. For a mixture of perfect gases at two temperatures $C_1 = C_2 = C$, so that we may let:

$$K_1/C = D_1 \tag{A2.9}$$

$$K_2/C = D_2 \tag{A2.10}$$

$$b/C = \mu \tag{A2.11}$$

Solutions to equations A2.7 and A2.8 may be derived by substitutions of the form:

$$T_1 = \alpha_1 \exp(ikx) \exp(\theta t) \tag{A2.12}$$

$$T_2 = \alpha_2 \exp(ikx) \exp(\theta t) \tag{A2.13}$$

whence:

$$\theta = -\mu - \frac{(D_1 + D_2)}{2} k^2 \pm \sqrt{\left\{ \left(\frac{D_1 - D_2}{2}\right)^2 + \mu^2 \right\}} \tag{A2.14}$$

which determines θ as a function of the Fourier transform variable k. For an instantaneous line source produced initially entirely in the electron gas, corresponding to a fission fragment track with uniform heating along its length, the solution for the lattice temperature T_2 becomes:

$$T_2 = \frac{Q_1}{2\pi C_1} \exp(-\mu t) \int_0^\infty \exp(-(D_1 + D_2)k^2 t/2) \frac{\sinh(\mu t \sqrt{(1 + \delta^2)})}{\sqrt{(1 + \delta^2)}} J_0(kr)k\, dk \tag{A2.15}$$

where:

$$\delta = \frac{(D_1 - D_2)}{2\mu} k^2 \tag{A2.16}$$

and Q_1 is the heat generated per unit length of the source.

References

OZEROFF, J. (1949), AEC Research Report No. AECD-2973.
BROOKS, H. (1950), AEC Research Report No. TID-10033.

The Ranges of Secondary Electrons in the Lattice

It is assumed that the secondary electron essentially follows a random Brownian-motion path through the crystal before being reduced to the general background energy through collision processes. It is also assumed that only interactions with lattice vibrations are permitted. This then allows us to draw on a theory due to Dekker (1954, 1958) who has considered the case of the temperature dependence of the secondary electron emission from crystals under primary bombardment. The average energy loss of the secondary in each interaction is independent of the initial energy and is only a function of temperature (Seitz, 1948, 1949). Denoting the average energy loss per collision by $\alpha(T)$, the energy of the secondary as a function of the number of collisions n it has suffered since it was energized by a δ-particle is given by:

$$E(n) = E_s - n\alpha(T) \qquad (A3.1)$$

Also, the mean-free-path for collisions with lattice vibrations for electrons with several electron volts of energy is proportional to the energy multiplied by a function of the temperature $\beta(T)$. We may therefore write for the mean-free-path:

$$\lambda(E, T) = \lambda_0 E_\beta(T) \qquad (A3.2)$$

where λ_0 is a constant. According to (A3.1) the energy decreases linearly with n, and therefore so does λ. The 'lifetime' of a secondary in falling to an energy E' is hence limited to a maximum number of scattering collisions n_m such that

$$n_m \alpha(T) = E_s - E' \qquad (A3.3)$$

If we now radically simplify the problem further and assume that the Brownian-motion is one-dimensional we may write the mean square displacement of a secondary during its 'lifetime' of n_m collisions as follows:

$$\langle x^2 \rangle_{av} = n_m \langle \lambda^2 \rangle_{av} = n_m \lambda_0^2 [\beta(T)]^2 \langle E^2 \rangle_{av} \qquad (A3.4)$$

where the averages must be taken over the n_m collisions. The quantity $\langle x^2 \rangle_{av}$ may then be considered as the square of a secondary range x_s'. Now according to (A3.1) we may write:

$$\langle E^2 \rangle_{av} = E_s^2 + \alpha^2 \langle n^2 \rangle_{av} - 2\alpha E_s \langle n \rangle_{av} \qquad (A3.5)$$

247

The average value of n is simply $n_m/2$, and if $n_m \gg 1$ we have:

$$\langle n^2 \rangle_{av} = \int_0^{n_m} \frac{n^2}{n_m} \, dn = \frac{n_m^2}{3} \tag{A3.6}$$

Making use of (A3.3) one readily finds that $\langle E^2 \rangle_{av}$ is independent of temperature and is only determined by the constants E_s and E':

$$\langle E^2 \rangle_{av} = \tfrac{1}{3}(E_s^2 + E'^2 + E_s E') \tag{A3.7}$$

Hence the mean square displacement may be expressed by:

$$\langle x^2 \rangle_{av} = \frac{\lambda_0^2}{3} (E_s - E')(E_s^2 + E'^2 + E_s E') \frac{[\beta(T)]^2}{\alpha(T)} \tag{A3.8}$$

and the *incremental effective range* of a secondary electron on this simple model becomes:

$$x_s' = k\beta(T)/[\alpha(T)]^{1/2} \tag{A3.9}$$

where k is a constant factor.

References

DEKKER, A. J. (1954), *Phys. Rev.* **94**, 1179.
DEKKER, A. J. (1958), *Solid State Physics* **6**, 251.
SEITZ, F. (1948), *Phys. Rev.* **73**, 550.
SEITZ, F. (1949), *Phys. Rev.* **76**, 1376.

Interatomic Potentials for Gold and Copper

In order to make quantitative predictions on the form of collision cascades in metals, it is essential to know the nature of the forces between ions. Assuming that these forces are central and velocity independent we may write for the vector force F_{12} on ion 1 due to another ion 2.

$$\mathbf{F}_{12} = -\frac{\mathbf{r}_{2\cdot1}}{r_{12}}\frac{\mathrm{d}V_{12}(r_{12})}{\mathrm{d}r_{12}} \qquad (A4.1)$$

where V_{12} is the interaction potential between ions 1 and 2, and depends on only their separation, r_{12}. Since we are basically concerned with interaction energies greater than 10 eV we may neglect all forces due to the crystalline structure (cohesive energies are of the order of 4 eV), and F_{12} is always repulsive. Moreover, in this energy range, the interaction potential is much greater than the Coulombic repulsion e^2/r between two singly charged ions, so that $V_{12}(r)$ applies equally well to atom–atom and atom–ion interactions. In general we refer to the interaction potential as the interatomic potential, although we could also speak of an interionic potential. A good knowledge of the inter-atomic potential is essential to calculate energy loss cross-sections for energetic displaced ions, the directional dependence of the displacement energy, and the critical energies for focusing, channelling, and spike production.

Two factors contribute to the repulsive forces between ions – the electrostatic repulsion of the positive nuclei, and a repulsion arising from the overlap of closed electron shells in the two ions.

The Screened Coulomb Potential

Bohr suggested in 1948 that the nuclear repulsion could be represented by a screened Coulomb potential, V_{sc}; i.e. the Coulombic repulsion between the two nuclei modified by a factor $\exp(-r/a)$ resulting from the partial screening of the nuclei by their own inner electrons.

$$V_{\mathrm{sc}}(r) = \frac{Z_1 Z_2 e^2}{r}\exp(-r/a) \qquad (A4.2)$$

where the screening radius a is given by:

$$a = \frac{a_0}{(Z_1^{2/3} + Z_2^{2/3})^{1/2}} \tag{A4.3}$$

and a_0 is the radius of the first Bohr orbital.

The screening function $\exp(-r/a)$ is derived by comparison with the Thomas-Fermi statistical model of a single atom in which the electrostatic potential around the nucleus is given by:

$$V(r) = \frac{Ze}{r}\,\phi(r/a) \tag{A4.4}$$

where $\phi(r/a)$ is the Thomas-Fermi function and:

$$a = a_0/Z^{1/3} \tag{A4.5}$$

For r less than $2a$, $\phi(r/a)$ may be approximated by $\exp(-r/a)$. Thus $V_{sc}(r)$ given by equation A4.2 should accurately represent the nuclear repulsion for separations less than $2a$. For gold this corresponds to $r \sim 0.18$ Å, and interaction energies greater than 60 keV: for copper, $r \sim 0.25$ Å, and energies greater than 7 keV. Comparison with a more accurate potential due to Abrahamson and colleagues (1961) based on Thomas-Fermi statistics for two ion systems shows in fact that, for copper, equation A4.2 is a good approximation for energies greater than 4 keV.

The Born-Mayer Potential

Born and Mayer proposed in 1932 that in ionic crystals the potential due to closed electron shell overlap could be described by a pure exponential of the form:

$$V(r) = \Psi \exp[\,(r_1 + r_2 - r)/\rho\,] \tag{A4.6}$$

where Ψ is a parameter depending on the ions considered, r_1 and r_2 are their ionic radii, and ρ is a constant, 0.345 Å. Later workers have found that, in metals, better agreement with experimental results can be obtained by considering both Ψ and ρ as variable parameters. We write the Born-Mayer potential in the form:

$$V_{BM}(r) = A \exp(-Br) \tag{A4.7}$$

Although the Born-Mayer potential is largely empirical, some theoretical justification is found in the radial charge densities in metal ions. In Figure 109 we plot the total charge outside a sphere of radius r in Al^{3+}, Cu^+, Ag^+, and Au^+ as determined by self-consistent-field solutions for the electron orbital functions (Hartree *et al.*, 1935, 1936; Douglas *et al.*, 1955; Pratt, 1952). In all four cases the exterior charge in the outer shells decreases exponentially with radius. The results are summarized in Table XIII, assuming that the exterior charge, $\xi(r)$, equals $\xi_0 \exp(-Br)$. Without finding solutions for the case of two ions, we intuitively expect that the degree of closed shell overlap and hence the repulsive potential should also show an exponential decrease with interatomic distance. In both copper and gold the exponential relationship holds down to a radius of about 0·5 atomic units, so that the Born-Mayer potential should describe the closed shell repulsion for separations greater than 1 atomic unit ($\sim 0·5$ Å). Using our eventual values of A and B this corresponds to interaction energies of less than 1·6 keV and 40 keV for copper and gold respectively.

TABLE XIII

Exterior charge for various metallic ions

$$\xi(r) = \xi_0 \exp(-\beta r)$$

Ion	Z	ξ_0	$\beta(\text{Å}^{-1})$	$\xi(r_0/2)$
Al^{3+}	13	36	6·1	0·001
Cu^+	29	48	4·4	0·13
Ag^+	47	83	3·9	0·34
Au^+	79	64	2·9	1·00

The Born-Mayer Potential for Gold

The following properties in metals are strongly dependent on the ionic repulsion:

(1) The elastic moduli.
(2) The rate of change of elastic moduli with pressure.
(3) The lattice constant.
(4) The cohesive energy.

Several methods for deriving values of A and B for copper, silver, and gold using the above properties are given by Mann and Seeger (1960).

However, all the above properties depend on the ionic repulsion at or near the interatomic separation, and extrapolation back to separations of 1 or 2 Å is not always valid. For example, Huntington (1953) has calculated A and B for copper using the three elastic moduli corrected for the effect of the free electrons and the electrostatic forces in a periodic lattice. With $A = 1·2 \times 10^5$ eV and $B = 5·73$ Å^{-1} the three elastic moduli are predicted with a mean deviation of only about 2%. Applying the same method to silver and gold, it is impossible to get a good correlation between the two dependent parameters, A and B, and the three independent moduli. The best mean deviations possible are 10% for silver and 50% for gold.

From Figure 109 and Table XIII we can see why this is so. Huntington's model assumes that the ionic repulsions are purely central (i.e. dependent only on separation). Table XIII shows that in the free state copper, silver, and gold ions have total charges of 0·13, 0·34, and 1·00 respectively outside a sphere of radius $r_0/2$ (half the nearest neighbour distance in the lattice). Hence the crystal structure must have a far greater effect on the outer electron shell in gold than in copper, with silver an intermediate case. Whereas in crystalline copper the outer 3d shell retains an almost spherically symmetric charge distribution, the outer 5d shell in gold does not. Therefore in gold the ionic repulsion near the interatomic separation should be strongly directional, and extrapolation back to smaller separations where the repulsion becomes purely central is quite

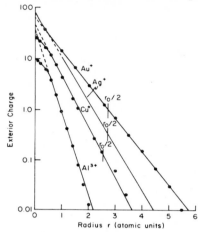

Figure 109. The total charge, $\xi(r)$, outside a sphere of radius r (in atomic units) for Al^{3+}, Cu^+, Ag^+ and Au^+. $r_0/2$, half the equilibrium separation in the crystal lattice, is indicated for Cu^+, Ag^+, and Au^+. Various parameters associated with these curves are listed in Table XIII.

difficult. This conclusion is in agreement with the results of Daniels and Smith (1958) who found that the pressure dependence of the sheer moduli in silver and gold could not be explained by central ion-repulsive forces.

If we could measure the repulsive potential $V_L(r)$ of two nearest neighbours in a gold lattice along the line joining their centres we might expect a potential

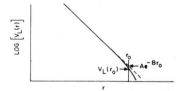

Figure 110. A schematic representation of the interatomic repulsive potential. $V_{L(r)}$, plotted logarithmically between two nearest neighbours in a gold lattice measured along the line joining their centres.

of the form shown schematically in Figure 110. For $r \ll r_0$, $V_L(r)$ equals $V_{BM}(r)$; near equilibrium the non-spherical charge distribution decreases $V_L(r)$ below $V_{BM}(r)$, and the slope of the logarithmic curve must be greater than B. Thus $V_L(r_0)$ gives an underestimate of $V_{BM}(r_0) = A\exp(-Br_0)$, while

$$-\left.\frac{\partial \log_e V_L(r)}{\partial r}\right|_{r_0}$$

gives an overestimate of B. These may be found from the bulk modulus and the rate of change of bulk modulus with pressure, since, under conditions of pure compression, non-central forces have no effect. The pertinent equations for fcc metals are (Daniels and Smith, 1958):

$$\frac{C_{11} + 2C_{12}}{3} = \frac{2N}{3}\left[r^2 \frac{\partial^2 V_L}{\partial r^2}\right]_{r_0} + \frac{2N'}{3}\epsilon_F \tag{A4.8}$$

$$\frac{d(C_{11} + 2C_{12})/3}{d\log_e r} = \frac{2N}{3}\left[r^2 \frac{\partial^3 V_L}{\partial r^3} - 3r^2 \frac{\partial^2 V}{\partial r^2}\right]_{r_0} + \frac{14}{3} N'\epsilon_F \tag{A4.9}$$

where $(C_{11} + 2C_{12})/3$ is the bulk modulus, N is the atomic density, r_0 is the nearest neighbour distance, N' is the free electron density, and ϵ_F is the Fermi energy. Taking $(C_{11} + 2C_{12})/3 = 18\cdot03 \times 10^{11}$ dyn/cm^2 (Neighbours and Alers, 1958)

$$\frac{d(C_{11} + 2C_{12})/3}{d\log_e r} = 320 \times 10^{11} \text{ dyn/cm}^2 \text{ (Daniels and Smith, 1958)}$$

$$N' = N = 5\cdot87 \times 10^{22}/\text{cm}^3$$
$$\epsilon_F = 5\cdot51 \text{ eV}$$

and assuming
$V_L(r) = \alpha \exp(-Br)$ for $r \sim r_0$, we arrive at the values
$$\alpha = 7\cdot8 \times 10^5 \text{ eV}$$
$$\beta = 5\cdot48 \text{ Å}^{-1}$$
$$V_L(r_0) = 0\cdot11 \text{ eV}$$

Hence $B < 5\cdot48$ Å^{-1} and $A \exp(-Br_0) > 0\cdot11$ eV.

To determine accurate values of A and B we must have some knowledge of $V(r)$ in the region 1 Å $< r < r_0$. The property best known theoretically and experimentally in this region is the maximum energy for the propagation of a $\langle 110 \rangle$ focusing sequence, E_f^{110}.† Thompson (1963) has found $E_f^{110} = 280 \pm 50$ eV, and a detailed theoretical treatment by Lehmann and Leibfried (1961) shows that E_f^{110} may be found by a solution for the equation:

$$2 = \frac{D^{110}}{2R} + 0\cdot347 \frac{D^{110}}{R} \cdot \frac{1}{BR} - \frac{0\cdot48 + 0\cdot17\, D^{110}/R}{1 - 2R/D^{110}}\left(\frac{1}{BR}\right)^2 \tag{A4.10}$$

where D^{110} is the interatomic separation in a $\langle 110 \rangle$ direction and

$$R = (1/2B) \ln(2A/E) \tag{A4.11}$$

Solving this equation for A and B and bearing in mind the estimates found above we arrive at the 'best' values:

$$A = 3\cdot6 \times 10^5 \text{ eV}$$
$$B = 5\cdot0 \text{ Å}^{-1}$$

† Although the displacement energy E_d is known with greater accuracy than E_f^{110} its relationship to $V(r)$ is not so clear.

For the sake of comparison we list in Table XIV the values of A and B used by other workers for the Born-Mayer potential in gold and some of their predicted results. $E_p^{100} = 4A \exp(-Br_0/\sqrt{2})$ is the potential barrier set up by the ring of four nearest atoms about a $\langle 100 \rangle$ direction. Since the direction of easiest displacement in gold is probably $\langle 100 \rangle$ (Bauer and Sosin, 1964), E_p^{100} is a very rough estimate of the displacement energy E_d^{100}. It is of necessity an over-estimate since it does not take into account relaxation of the ring atoms as the displaced atom passes through. Experimentally it is found that $E_d^{100} = 35$ eV (Bauer and Sosin, 1964).

We note that while the stiff potentials (large B) due to Seeger (1962) and van Jan (1965) give reasonable values of E_f^{110} at $r \sim r_0/2$, they decrease too rapidly with increasing r, and the values of E_p^{100} at $r \sim r_0/\sqrt{2}$ and $V(r)$ at $r \sim r_0$ appear to be too small. This lends qualitative support to the softer potential derived here. The very different potential of Anderson and Sigmund

TABLE XIV
Born-Mayer potential, $A \exp(-Br)$, for gold

Author	A (eV)	B (Å^{-1})	$V(r_0)$ (eV)	E_f^{110} (eV)	E_p^{100} (eV)
Seeger (1962)	$4 \cdot 4 \times 10^5$	$5 \cdot 6$	$0 \cdot 04$	140	20
	15×10^5	$5 \cdot 75$	$0 \cdot 088$	370	50
	119×10^5	$6 \cdot 95$	$0 \cdot 022$	440	35
van Jan (1965)	11×10^5	$6 \cdot 0$	$0 \cdot 033$	175	22
Thompson and Nelson (1961)	8×10^5	$5 \cdot 22$	$0 \cdot 24$	430	79
Brinkman II (1960)	$6 \cdot 7 \times 10^5$	$5 \cdot 42$	$0 \cdot 11$	280	45
Andersen and Sigmund (1965)	$3 \cdot 7 \times 10^4$	$4 \cdot 57$	$0 \cdot 07$	55	14
This work	$3 \cdot 6 \times 10^5$	$5 \cdot 0$	$0 \cdot 20$	280	56

(1965) gives values of E_f^{110}, E_p^{100}, and $V(r_0)$ which all appear to be much too small. Although they based their values of A and B on the known focusing and displacement energies they used different criteria to relate E_d^{100} and E_f^{110} to $V(r)$. The small value of $V(r_0)$, however, is less open to different interpretations, and does not lend much support to their values. The Brinkman II (1960) and Thompson-Nelson (1961) potentials give good results for E_p^{100}, E_f^{110}, and $V(r_0)$ and qualitatively are not too different from ours. As is discussed below the agreement with Brinkman's potential is probably fortuitous since the general Brinkman II potential predicts a dependence of B on Z which does not appear to be justified. The Thompson-Nelson values were found by comparison with experimental focusing data and the bulk modulus, so the agreement is not unexpected.

Because of our methods of derivation, our potential appears to behave excellently at $r \sim r_0/2$ (E_f^{110}) and very reasonably near $r \sim r_0$. Moreover at

$r \sim r_0/\sqrt{2}$, $E_p^{100}/E_d^{100} \sim 1\cdot6$ is not an unreasonable figure for the relaxation effects mentioned above. We may use our values of A and B for V_{BM} with some confidence for separations up to r_0 and energies greater than $0\cdot2$ eV.

As discussed above there should be some maximum interaction energy above which the Born-Mayer potential does not accurately represent the closed shell repulsion. We can estimate this energy by comparing experimentally determined ranges of gold ions in a gold lattice with theoretical ranges based on a Born-Mayer repulsion. The only available data on the range of gold ions in gold is that of van Lint *et al.* (1961) for the energy range 6 to 14 keV. Theoretical ranges can be determined analytically by assuming that the crystal consists of a random array of atoms, and that the scattering of the incident ion is governed by hard sphere collisions (Holmes and Leibfried,

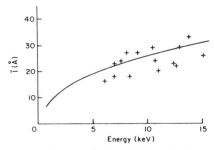

Figure 111. The mean total path length, l, of a gold ion slowing down in a gold lattice from energy E to the displacement energy ($E_d = 35$ eV) for the Born-Mayer interatomic potential derived in this work. The crosses are the experimental work of van Lint and his colleagues (1961).

1960). The mean total path length, $\bar{l}$, travelled by an ion in slowing down from an energy E to the displacement energy E_d is:

$$\bar{l} = \frac{B^2}{\pi N} \left[\frac{1}{\log_e^2 (2A/E)} + \frac{1}{\log_e(2A/E)} - \frac{1}{\log_e(2A/E_d)} \right] \qquad \text{(A4.12)}$$

In Figure 111 we plot $\bar{l}$ for the values of A and B derived above, and give the experimental results of van Lint *et al.* The agreement is good. The actual mean range and rms range will be somewhat less than $\bar{l}$, but the inherent error in both the model and the experimental results do not justify more detailed calculations. We conclude that our Born-Mayer potential for gold is valid for interaction energies up to at least 15 keV, or for interatomic separations greater than $0\cdot6$ Å. This compares well with the theoretical limits of 40 keV and $0\cdot5$ Å derived previously.

The Screened Coulomb Potential for Gold

It has been suggested by some workers that the screened Coulomb potential as initially suggested by Bohr (1948) be modified so that the screening radius a

is given by:

$$a = \lambda \frac{a_0}{(Z_1^{2/3} + Z_2^{2/3})^{1/2}},$$
(A4.13)

where λ is an adjustable parameter. Analysing van Lint's results for ranges in gold between 6 and 15 keV, Holmes (1962) found a value of $\lambda = 2\cdot3$ gave the best agreement. However we have shown above that van Lint's results can be adequately explained by the use of a Born-Mayer potential, and the closed shell repulsion, not the nuclear repulsion, should be the dominant force in this energy region. Unfortunately no range measurements of gold ions in gold at

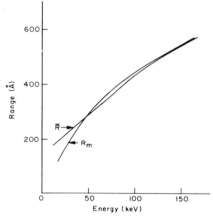

Figure 112. The mean penetration depth, R_m, of gold ions in gold extrapolated from the results of Uhler *et al.* (1963) on tungsten, and the theoretical rms range, $\bar{R}$, determined for the screened Coulomb potential.

higher energies are available. The penetration depths of W^{187} ions in tungsten have been measured by Uhler *et al.* (1963) for energies less than 127 keV. The small differences in mass number (184 *v.* 197), atomic number (74 *v.* 79) and density (19·3 in both cases) between tungsten and gold suggest that results in the two metals should be quite close. We have re-interpreted Uhler's results in terms of gold by comparison with the range-energy formulae of Lindhard, Scharff, and Schiøtt (1963), and plotted the mean penetration depth in Figure 112. We also plot the rms range $\bar{R}$ given by Holmes and Leibfried (1960) for the screened Coulomb potential for gold with $\lambda = 1$. The agreement is remarkably good for $E > 50$ keV. This is possibly fortuitous. The important point is that, since $\bar{R} \propto 1/\lambda^2$, taking $\lambda = 2\cdot3$ reduces $\bar{R}$ by a factor of 5, thus completely destroying the agreement between theory and experiment. The conclusion must be that the screened Coulomb potential as originally proposed by Bohr is valid in gold for interaction energies greater than 50 keV.

In trying to bridge the gap between the Born-Mayer and screened Coulomb potentials we run into difficulties. With our choice of parameters $V_{BM} > V_{SC}$ for energies less than 200 keV, and the ranges predicted using V_{BM} for $20 < E < 150$ keV are an order of magnitude lower than those extrapolated from Uhler's results on tungsten. On the other hand the extrapolated gold ranges from Uhler's work are several times greater than van Lint's ranges for $6 < E < 15$ keV, and it is difficult to assess where the error lies. It appears, however, that extrapolation of V_{BM} above 20 keV to describe closed shell overlap is not justified. The best solution for the total interaction potential appears to be some sort of numerical interpolation between V_{BM} at low energies and V_{SC} at high energies.

In summary we find that for interaction energies less than 15 keV or for interatomic separations greater than 0·6 Å the interatomic potential in gold is best represented by the Born-Mayer potential:

$$V_{BM}(r) = A \exp(-Br) \tag{A4.7}$$

with

$$A = 3\cdot6 \times 10^5 \text{ eV}$$
$$B = 5\cdot0 \text{ Å}^{-1}$$

while for interaction energies greater than 50 keV or for interatomic separations less than 0·2 Å it is best represented by the screened Coulomb potential:

$$V_{SC}(r) = Z^2 \exp(-r/a)/r \tag{A4.14}$$

with

$$Z^2 = 9\cdot0 \times 10^4 \text{ eV Å}$$
$$a = 0\cdot087 \text{ Å}$$

For the intermediate ranges a numerical interpolation appears to be necessary.

The Interatomic Potential for Copper

For copper we take as an effective interatomic potential the sum of the Born-Mayer and screened Coulomb potentials as proposed by Leibfried (1960):

$$V(r) = A \exp(-Br) + \frac{Z^2}{r} \exp(-r/a) \tag{A4.15}$$

and

$$A = 2\cdot2 \times 10^4 \text{ eV}$$
$$B = 5\cdot1 \text{ Å}^{-1}$$
$$Z^2 = 1\cdot2 \times 10^4 \text{ eV Å}$$
$$a = 0\cdot12 \text{ Å}$$

The values of A and B are those of the Gibson II potential (1960) which has proved successful in past calculations, and a is as given by Bohr (1948).

We have already noted that in copper V_{BM} should be valid for $E < 1\cdot6$ keV and V_{SC} valid for $E > 7$ keV. Since the cross-over point where $B_{BM} = V_{SC}$, 6 keV, lies in the intermediate region where neither is strictly valid, the addition of the two gives a potential which approximates to V_{BM} or V_{SC} in the appropriate energy ranges. Moreover it has the advantage of being analytic in the intermediate region. Such a simple approach is not possible for gold since the cross-over point falls well outside the intermediate region.

The Dependence of B on the Atomic Number

Brinkman (1960) has proposed an empirical interaction potential (the Brinkman II potential)

$$V(r) = AZ_1 Z_2 e^2 \frac{\exp(-B'r)}{1 - \exp(-Ar)} \qquad (A4.16)$$

where

$$B' = \frac{(Z_1 Z_2)^{1/6}}{1 \cdot 5\, a_0} \qquad (A4.17)$$

$$A = 2 \cdot 58 \times 10^{-6}(Z_1 Z_2)^{7/4}\, \text{Å}^{-1}$$

For small r this approaches the screened Coulomb, $Z_1 Z_2 e^2 \exp(-R'r)$; and at large r the Born-Mayer $AZ_1 Z_2 \exp(-B'r)$. In both cases the exponential factor $\exp(-B'r)$ is the same. However the exponentials in the two potentials arise from two separate causes, the screening of the nuclei by the *inner* electrons for the screened Coulomb, and the overlap of the *outer* electron shells for the Born-Mayer. For the former Bohr has proposed a factor $\exp(-B''r)$ where $B'' < Z^{1/3}$ in agreement with Brinkman's B'. In the above we assumed that the exponential behaviour of the Born-Mayer potential, $A\exp(-Br)$, and that of the exterior charge $\xi_0 \exp(-\beta r)$, were somehow related. Table XIII shows that β definitely decreases with increasing Z, and B should behave similarly. Thus the best value of B for gold, $5 \cdot 0$, is less than that for copper, $5 \cdot 1$, although the difference is small. Since Brinkman's B' increases with increasing Z (a factor of $1 \cdot 4$ in going from copper to gold), its presumed equivalence to B is possibly not well justified, and we conclude that the Born-Mayer limit of the Brinkman II potential is not necessarily accurate. Similar conclusions have been reached by Andersen and Sigmund (1965).

References

ABRAHAMSON, A. A., HATCHER, R. D., and VINEYARD, G. A. (1961), *Phys. Rev.* **121**, 159.

ANDERSEN, H. H., and SIGMUND, P. (1965), Risö Report No. 103.

BAUER, W., and SOSIN, A. (1961), *Phys. Rev.* **136**, 255.

BOHR, N. (1948), *Met. Fys. Medd. Dan. Vid. Selsk.* **18**, 8.

BORN, M., and MAYER, L. (1932), *Z. Physik.* **75**, 1.

BRINKMAN, J. A. (1960), Rendiconti della Sinola Internazionale di Fisica 'E. Fermi', XVIII Corso, 830.

DANIELS, W. B., and SMITH, C. B. (1958), *Phys. Rev.* **111**, 713.

DOUGLAS, A. S., HARTREE, D. R., and RUNCIMAN, W. A. (1955), *Proc. Camb. Phil. Soc.* **51**, 486.

GIBSON, J. B., GOLAND, A. N., MILGRAM, N., and VINEYARD, G. A. (1960), *Phys. Rev.* **120**, 1228.

HARTREE, D. R. (1935), *Proc. Roy. Soc. A* **151**, 96.

HARTREE, D. R., and HARTREE, W. (1936), *Proc. Roy. Soc. A* **157**, 480.

HOLMES, D. K. (1962), *Radiation Damage in Solids*, IAEA (Vienna), **1**, 3.

HOLMES, D. K., and LEIBFRIED, G. (1960), *J. Appl. Phys.* **31**, 1046.

HUNTINGTON, H. B. (1953), *Phys. Rev.* **91**, 1092.

JAN, R. VAN (1965), *Phys Stat. Sol.* **8**, 331.

LEHMANN, C., and LEIBFRIED, G. (1961), *Z. Physik.* **162**, 203.

LEIBFRIED, G. (1960), Rendiconti della Sinola Internazionale di Fisica 'E. Fermi', XVIII Corso, 227.

LINDHARD, J., SCHARFF, M., and SCHIØTT, H. E. (1963), *Math.-fys. Medd.* **33**, 14.

LINT, V. A. J. VAN, SCHMITT, R. A., and SUFFREDINI, C. A. (1961), *Phys. Rev.* **121**, 1457.

MANN, E., and SEEGER, A. (1960), *J. Phys. Chem. Sol.* **12**, 314.

NEIGHBOURS, J. R., and ALERS, G. A. (1958), *Phys. Rev.* **111**, 707.

PRATT, G. W. (1952), *Phys. Rev.* **88**, 1217.

SEEGER, A. (1962), *Radiation Damage in Solids*, IAEA (Vienna), **1**, 101.

THOMPSON, M. W. (1963), *Phys. Letters* **6**, 24.

THOMPSON, M. W., and NELSON, R. S. (1961), *Proc. Roy. Soc. A* **259**, 458.

UHLER, J., DOMEIJ, B., and BORG, S. (1963), *Arkiv. Fysik.* **24**, 413.

Index

DATE DUE